SUBNUCLEAR PHYSICS

THE FIRST 50 YEARS:
HIGHLIGHTS FROM ERICE TO ELN

World Scientific Series in 20th Century Physics

Forthcoming

World Scientific Series in 20th Century Physics – Vol. 24

SUBNUCLEAR PHYSICS

THE FIRST 50 YEARS:
HIGHLIGHTS FROM ERICE TO ELN

Antonino Zichichi

University of Bologna, Italy
National Institute of Nuclear Physics, Italy
CERN, Geneva, Switzerland
World Federation of Scientists, Beijing, Geneva, Moscow, New York

Edited by

O. Barnabei
Bologna Academy of Sciences

P. Pupillo
University of Bologna

F. Roversi Monaco
University of Bologna

This volume is an extended version of several lectures presented at:
The bicentenary celebrations of Luigi Galvani: Bologna Academy of Sciences, 7 March 1998, opening lecture
The centenary of the electron discovery: INFN–Frascati National Lab, 12 November 1997, invited lecture
The centenary of the Italian Physical Society: Como, 28 October 1997, invited lecture
The Subnuclear Physics 50th Anniversary: Erice, 2 September 1997; INFN–Gran Sasso Lab—L'Aquila, 8 May 1998

World Scientific
Singapore • New Jersey • London • Hong Kong

Published by

World Scientific Publishing Co. Pte. Ltd.

P O Box 128, Farrer Road, Singapore 912805

USA office: Suite 1B, 1060 Main Street, River Edge, NJ 07661

UK office: 57 Shelton Street, Covent Garden, London WC2H 9HE

Library of Congress Cataloging-in-Publication Data
Zichichi, Antonino.
   Subnuclear physics : the first 50 years : highlights from Erice to ELN / Antonino
Zichichi ; edited by O. Barnabei, P. Pupillo, F. Roversi Monaco.
    p. cm. -- (World Scientific series in 20th century physics ; vol. 24)
   Includes bibliographical references and index.
   ISBN
   1. Particles (Nuclear physics) I. Barnabei, O. (Ottavio) II. Pupillo, P. (Paolo) III.
Roversi Monaco, Fabio A. IV. Title. V. Series.

  QC793.28.Z53   1999
  539.7'2--dc21   99-048252

British Library Cataloguing-in-Publication Data
A catalogue record for this book is available from the British Library.

Photos from the Dirac Museum, selected by Norma Sanchez, Director of the Museum.

First edition jointly published in 1998 by the Bologna Academy of Sciences and the University of Bologna.

This edition copyright © 2000 by World Scientific Publishing Co. Pte. Ltd.

Printed in Singapore.

«There is properly no history; only biography.»

Ralph Waldo Emerson

CONTENTS

FOREWORD

Fifty years of research using protons, electrons, neutrons and nuclei, in the energy range from a few tens up to a few hundreds GeV, have given rise to the most powerful synthesis of all time in our understanding of nature. Here are the basic concepts and achievements. The origin of all fundamental forces of nature is the same: the Gauge Principle. There are three families of fundamental building blocks of matter (quarks and leptons). When viewed using the Fermi forces, which provide the fuel for all the stars and control their rate of burning, quarks and leptons appear mixed. All basic quantities vary with energy. Not only real but imaginary masses exist. The existence of imaginary masses has an important effect on our description of the vacuum. In turn, the vacuum determines the masses and the interaction properties of the real particles. Neutrinos appear not to be massless. All gauge couplings seem to converge towards the same value at an energy level two orders of magnitude below the maximum conceivable energy, the Planck scale. Here very intense new theoretical developments are taking place. In order to explain how it is possible to keep two energy scales separated, the Fermi and the Planck scales, which are seventeen orders of magnitude apart, Nature should have a symmetrical structure which puts fermions and bosons on equal grounds. The point-like description of our world must be abandoned in favour of a non-point-like (string or membrane) elementary basic entity. Our four-dimensional space-time is the result of an expansion of hidden dimensions — of both bosonic $(10 + 1)$ and fermionic (32) nature — whose number could be as high as 43: this is the superspace. These are the main theoretical achievements which arose from the three experimental discoveries, all announced in the golden year 1947: the Lamb-shift, the π–meson and the V^0 particles. The logical connection of these three starting points, with the enormous developments of our way of thinking about natural phenomena, is the basis of this review.

The experimental results acquired during these past fifty years in subnuclear physics tell us that the Standard Model cannot be the definitive theory, in spite of the fact that it is the most powerful synthesis of all known and rigorously measured phenomena.

Looking back at the last fifty years, the amount of new knowledge acquired is really overwhelming. We could relax and enjoy the Standard Model, but we already know that this superb synthesis is just the starting point of a new horizon. For this new horizon to be investigated, a project for a new collider able to work at extreme energy and luminosity is needed. This is ELN.

The review of the main steps, from the year subnuclear physics was born (1947) to now, is presented as seen from my reference frame, and cannot therefore be unbiased. Let me cite Rabi: «*Physics is Intellectual Freedom. Our interest is to understand nature. It is to our liking to choose the best way. Every physicist has his own interests and his own likes and dislikes*».

This book is a personal reconstruction — since its birth — of the development of subnuclear physics where my understanding of the logic of Nature is associated with a concrete concern about the future of our field. It is this concern at the origin of my activity devoted towards the implementation of new projects, including the first School in Subnuclear Physics. Being personal, the reconstruction deals with some major issues which have attracted my own interest. Here they are.

— The problem of divergences in electro-weak interactions and the third lepton. Physics results are finite: why then do theoretical calculations, for example in Quantum Electro-Dynamics (QED), give rise to divergent quantities? High precision measurements in QED and in weak interactions attracted my interest. In fact, for example, the contributions of virtual weak processes in the elementary properties of the muon should produce divergent results. The cure for divergences was called, by theorists, renormalization. My career started with the first high precision measurements of the muon electromagnetic and weak basic properties, i.e. its anomalous magnetic moment (measured with $\pm\, 5 \times 10^{-3}$ accuracy), and its weak charge (measured with $\pm\, 5 \times 10^{-4}$ accuracy). The invention of new technologies for the detection of electrons and muons allowed me to start searching for a new lepton and to make a series of high precision QED tests. This activity culminated with the discovery by others of the third lepton (now called τ) to which I have devoted ten years of my life.

— The violation of the symmetry operators (C, P, T, CP) and Matter-Antimatter Symmetry. I was very much intrigued by the great crisis arising from the discovery that the basic invariance laws (parity, charge conjugation, time reversal) were not valid in some elementary processes. Together with the successes of the S-matrix and the proliferation of "elementary" particles, the powerful formalism called Relativistic Quantum Field Theory (RQFT) appeared to be in trouble. These were the years when the RQFTs were only Abelian and no-one was able to understand the nature of the strong forces. The discovery of CP breaking gave a new impetus to the search for the first example of nuclear antimatter: in fact the antideuteron had been searched for and found not to be produced at the 10^{-7} level. The order of magnitude of its production rate was unknown and finally found to be 10^{-8}. This level of detection was reached by my group thanks to the construction, at CERN, of the most intensive negative beam and of the most precise time-of-flight detector.

— The mixing in the pseudoscalar and vector mesons: the physics of Instantons. Another field thoroughly investigated by me is the mixing properties of the pseudoscalar and of the vector mesons, realized through the study of their rare decay modes. This could be accomplished thanks to the invention of a new detector, the neutron missing mass spectrometer. The physics issues could probably be synthetized in terms of the U(1) problem. In other words, why do the vector mesons not mix while the pseudoscalar mesons mix so much? This issue found — after many decades — a satisfactory answer when it was discovered by G. 't Hooft how Instantons interact with the Dirac sea.

— The non-Abelian nature of the Interaction describing quarks, gluons and the Effective Energy. Another puzzle to me was the enormous variety of multihadronic final states produced in strong, electromagnetic and weak interactions. Why are these final states all different? This appeared to be in contrast with the order of magnitude reduction in the number of mesons and baryons obtained, first with the eightfold way by Gell-Mann and Ne'eman and then with the quark proposal by Gell-Mann and Zweig. The discovery of Scaling at SLAC and the non existence of quarks found by us at CERN using the ISR was finally understood in terms of quarks, gluons, and of their non-Abelian interaction. With the advent of quarks and gluons, the puzzle became even more intriguing since the multihadronic final states had to have the same origin. The introduction of the new quantity, "Effective Energy", allowed to clarify the problem even though the quantitative QCD description is still missing.

— The hobby: the Supersymmetry threshold and its problems.

The reconstruction of the first five decades of subnuclear physics has a strong link with these topics and with the "Ettore Majorana" School of Subnuclear Physics at Erice, a small town on the top of a mountain founded — according to the myth — by the son of Venus. Here, every year since 1963, the development of subnuclear physics has been recorded and the hottest topics of the moment registered as faithfully as possible in the discussion sessions of the Erice School. At this School, I have attempted to have as Lecturers the most active and authoritative members of the subnuclear physics community. Their ingenuity, their wisdom, their rigorous attempts to understand the constituents and the fundamental forces of nature are reported in the volumes of the Subnuclear Physics Series. The original ideas which have flourished during the past 50 years were the focus of an intense intellectual activity, both for theorists and for experimentalists, who have contributed to the lectures and to the discussion sessions of the Erice Subnuclear Physics School.

This is why I have chosen the Erice School to illustrate the development of this exciting field of physics. To help the reader interested in deepening an argument or a topic mentioned in this book, there is an analytic index of the main subjects with reference to the Erice Subnuclear Series, in the appendix.

My physics interests made me very much concerned about the future of subnuclear physics. This concern lies at the origin of my activity devoted to the implementation of new projects. As quoted in the title, I draw a line from Erice to the proton collider with the highest energy and luminosity which could be built with simple extrapolation of the presently known technologies: the ELN project. The ELN project is very ambitious but we should be encouraged by our previous experiences. In fact, the path leading from Erice to the ELN has already gone through the Gran Sasso project (now the largest and most powerful underground laboratory in the world), the LEP-white-book which allowed this great European venture to overcome the many difficulties that had blocked its implementation during many years, the HERA collider (now successfully running), and the roots of LHC, as for example the 5-metre diameter for the LEP tunnel, and the LAA-R&D project, implemented to find the original detector technologies needed for the new colliders. These past achievements in project realization are mentioned in order to corroborate my optimism and enthusiasm in encouraging new actions and new ideas for the future of Subnuclear Physics in Europe and in the world.

PREFACE

For the Galvani Bicentenary Celebrations, the University and the Academy of Sciences of Bologna singled out Subnuclear Physics to be the field of scientific research associated with this important event, as it would best illustrate, for the new generation of students, the challenge inherent to fundamental sciences.

Subnuclear Physics was born 50 years ago and has represented, ever since, the new frontiers of Galilean Science. In his opening lecture delivered on the first day of the new academic year, Professor Antonino Zichichi analytically reviewed the basic conceptual developments and main discoveries achieved in Subnuclear Physics, during these last 50 years. Given the importance of this field of fundamental research, we invited our colleague and friend to expand the contents of his opening lecture into a volume to be published as our contribution to the Galvani Bicentenary Celebrations.

We wish to thank Professor Norma Sanchez from the Observatoire de Paris and director of the Dirac Museum, for having provided pictures and selected statements delivered by eminent Erice lecturers which helped illustrate the history of the First Fifty Years of Subnuclear Physics.

Ottavio Barnabei
President of the Bologna Academy of Sciences

Paolo Pupillo
Chairman of the Science Faculty, University of Bologna

Fabio Roversi Monaco
Rector of the University of Bologna
President of the Committee for the Luigi Galvani Bicentenary Celebrations

Bologna, Galvani Celebrations 1998

Luigi Galvani's laboratory. Visible in this engraving is the electrical apparatus Luigi Galvani used to produce an electric cell with the fluids extracted from the body of a frog.

I – INTRODUCTION.

It is for me a great privilege to inaugurate the activities of the Academy of Sciences of the oldest University in the world with a lecture reviewing the achievements in subnuclear physics over the fifty years since its birth.

The choice of subnuclear physics as the topic for the opening lecture of the Galvani Bicentenary Celebrations is the best testimony of the support given by the University of Bologna and its Academy of Sciences to my field of research.

Luigi Galvani is the father of macroelectricity. All facilities built, the world over, for subnuclear physics could not exist without the developments of electromagnetism on a large, gigantic scale: the greatest subnuclear physics laboratory in the world, CERN, would not be there, neither would DESY (Hamburg), FERMILAB (Chicago), BNL (New York), SLAC (Stanford), where the most powerful accelerators using protons, electrons and nuclei are in operation.

Let us not forget the formidable idea of Luigi Galvani of putting together Copper and Zinc: the (Cu-Zn) junction. Volta's cell could not have been invented if a man had not had the idea of the (Cu-Zn) junction.

The development of electricity and magnetism in the XIX century, the Maxwell synthesis, which allowed mankind to realize that the enormous variety of optical phenomena was a further manifestation of electromagnetism, started with the (Cu-Zn) junction. All this looks obvious nowadays, but think of the enthusiasm which brought Lord Kelvin to say, in his opening lecture at a physics conference a hundred years ago, that with the Maxwell equations the basic conceptual understanding of physics was over: few details are left to be measured. Six months later J.J. Thomson discovered the negative electron.

During the succeeding three decades, three important developments took place: i) the discovery by Planck of his quantum of action; ii) the Einstein reinterpretation of the Lorentz-invariance of the Maxwell equations in terms of the relative property of space and time, not being as conceived by Newton (absolute and independent) but relative and so strongly linked that all physics had to be described in four dimensions; iii) the discovery that the electron has half-integer spin and its gyromagnetic ratio is not one (as for the orbital angular momenta) but two.

These developments culminated with one of the greatest achievements of mankind: the discovery by Paul Adrien Maurice Dirac of his equation. This opened a new horizon in physics: the virtual phenomena with all their rigorously predicted and measurable effects. All this is illustrated in Table 1.

Dirac at Erice.

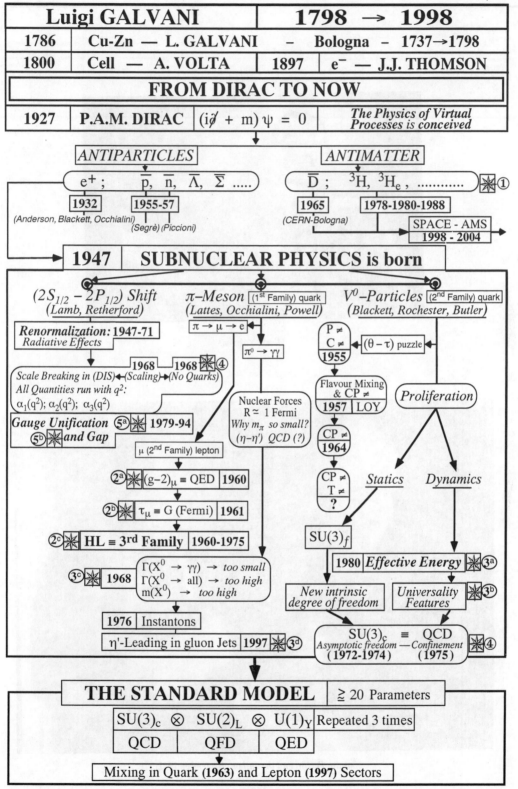

Table 1

Hideki Yukawa.

Sandro Pertini and the author, September 1982.

The great achievements of subnuclear physics could not have been accomplished without the existence of large facilities. This is the reason why I have devoted some of my thinking to this vital aspect of our field (§ I.1).

Subnuclear physics was born (see Table 1), when three discoveries were announced:

the Lamb-shift $\qquad$ [1],

the π–meson $\qquad$ [2],

the V–particles $\qquad$ [3].

These three great discoveries are now understood as being: 1) the first example of radiative effects [4]; 2) the first example of a bound system made of a quark-antiquark $(q\bar{q})$ pair; 3) the first example of a new flavour beyond the first family.

Before subnuclear physics was born there were big problems (§ I.2).

The lecture will consist of the following three main parts:

i)   Achievements in the past fifty years (§ II);

ii)   The ELN project (§ III);

iii)   Conclusions: Blackett and Russell (§ IV).

I.1   *Italy as an example.*

When, in 1979, I started to think about a new strategy for subnuclear physics to be promoted in Italy in order to bring its budget to a European level, the new Minister of Science and Technology (M. Pedini) had decided to ask (in a confidential but firm manner) the persons who had high responsibilities in governmental organizations to answer the following three questions:

1)   Have you contributed with original ideas to the development of your research field? If yes, which are the future developments you foresee?

2)   Have you ever had an original idea for a new project? If yes, have you been able to implement it?

3)   Is the project you are proposing in line with your past activity and competence in its scientific and technological implications?

The leading force in this unprecedented action aimed at restructuring Italian science was the new President of Italy, a glorious leader of antifascism, Sandro Pertini. I obeyed the instructions and that was the first step which finally brought INFN to an increase of budget by one order of magnitude. It was my first experience of interacting with decision-making leaders and I have never forgotten it.

On the occasion of the fifty years of subnuclear physics and of the promotion of the ELN project, I have followed the same guidelines. This is the reason why I have tried to do my best in sketching out the past fifty years of progress in subnuclear physics (§ II.1),

Works during the excavation of the Gran Sasso Laboratory halls.

in pointing out the role I have played in the field (§ II.2) and in reviewing the basic steps of the conceptual developments in subnuclear physics (§ II.3). I have tried to do my best to convince people that subnuclear physics has opened new horizons in the field of advanced human knowledge. This is the way in which the future of our science and technology can be guaranteed. This means strengthening the ELN strategy, started in 1979 when subnuclear physics in Italy was in a critical state, with a very low financial support and zero chance to implement new projects. The ELN strategy allowed the implementation of new projects in Italy and abroad.

As recently recalled [5], in Italy, the first proposal was to implement a project for the most powerful underground laboratory [6a, b, c, 7, 8, 9], whose aims were presented to a special committee of the Italian Senate (Fig. I.1.1). The basic characteristics of this new venture (reported in Fig. I.1.2) attracted the interest of the physics community world-wide.

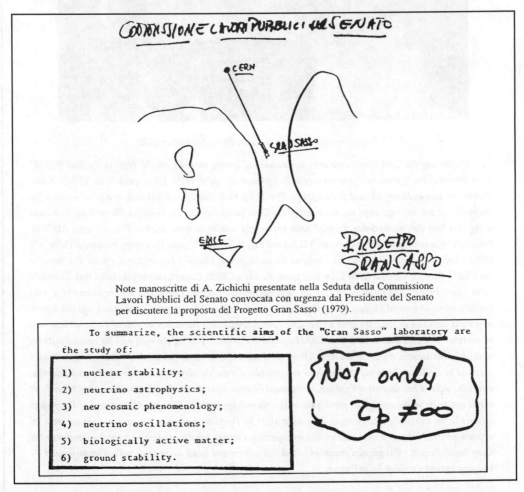

Fig. I.1.1: (Figure from Reference 5). In the upper part, a detail of the Gran Sasso project presented by A. Zichichi in the Public Work Committee of the Italian Senate. In the lower part the reproduction of page 13 of the original project [6a].

The author and Bruno Pontecorvo, Rome, September 1978.

During the Cold War Pontecorvo was accused of having passed from the West to the East (USSR) some secrets of the first nuclear weapon ever built by mankind. After many years spent in the USSR, Bruno Pontecorvo was suddenly allowed to visit Italy. The Berlin Wall was up and this became a great occasion for the media. It was the time when we had proposed the Gran Sasso Project. The so-called "Rome School" tried to bring to a halt the "Gran Sasso Project" with arguments similar to those used at Frascati with ADONE («Zichichi is searching for butterflies», see § II.2-2 and Fig. II.2-2.6) to boycott the energy increase of the (e^+e^-) collider (see § II.2-2 and Fig. II.2-2.3). Apparently, underground lobbying had produced effects. For instance the CERN-Director General (DG) Leon Van Hove, during a CERN Council meeting declared that Zichichi's Gran Sasso Project was invented to stop the new collaboration between Italy and France to realize a joint venture in underground physics using the Fréjus tunnel. After this unprecedented attack against a very important initiative in Italy, the other CERN-DG, John Adams, called the author into his office, to tell him «not to worry». This ended all attacks from CERN against the project. During the visit and the various lectures centred on Pontecorvo, a journalist asked: «Professor Pontecorvo, what do you think of the Gran Sasso Project proposed by Professor Zichichi? Many physicists consider it a useless Napoleonic venture with weak scientific content». After a few seconds of thinking, in the usual Pontecorvo style of soft and slow answering, he said: «I regret not to be young enough to participate in this formidable project. The scientific content of the project appears to me extremely interesting». This declaration by Pontecorvo came as a surprise, since we were on opposite political sides and every journalist was expecting a strong negative statement from Pontecorvo on the Gran Sasso Project. But physics prevailed. And this put an end to all — open as well as underground — lobbying against the Gran Sasso Project.

2. THE BASIC CHARACTERISTICS OF THE LAB.

The range of scientific perspectives opened up by the Gran Sasso Laboratory goes far beyond the measurement of the proton lifetime, as shown in Fig. 1.1.

These *scientific perspectives* depend on the *basic features* of the Gran Sasso Laboratory, which are:

1) very low noise due to local radioactivity;
2) neither too deep, nor too shallow underground;
3) orientation towards the most powerful (artificial) source of neutrinos and other unknowns (Fig. 2.1);
4) link with a laboratory at the top of the Gran Sasso, which allows time coincidences to be made (Fig. 2.2);
5) instrumentation which uses the most advanced technologies.

The *low noise* level in terms of *natural radioactivity*, was proved before the excavation work started. The measurements of the cosmic ray flux and of the local rock radioactivity were first performed by one of my collaborators, – L. Federici [3] – whom I want to pay tribute to on this solemn occasion. These measurements demonstrated that over the length of <u>one Km</u> the cosmic ray flux was constant. This nice feature is due to the shape and structure of the mountain. The Gran Sasso rock radioactivity was so low that the term ‹laboratory of cosmic silence› could be coined.

Reproduction of page 111 of Ref. 9.

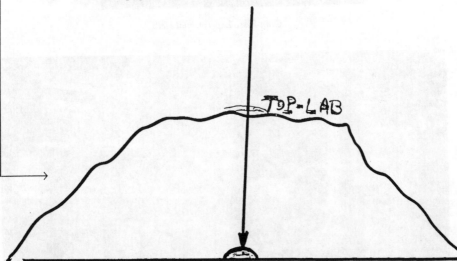

Side view section of the Gran Sasso mountain (hand-drawn by the author), showing the Underground Laboratory and the Top-Lab. Notice the "flat" shape of the mountain, which is at the origin of the constant cosmic ray flux over one kilometre along the tunnel. This drawing, together with the other on the previous page, was presented at the special meeting organized by the President of the Senate, H.E. Prof. A. Fanfani.

Fig. I.1.2: (Figure from Reference 5).

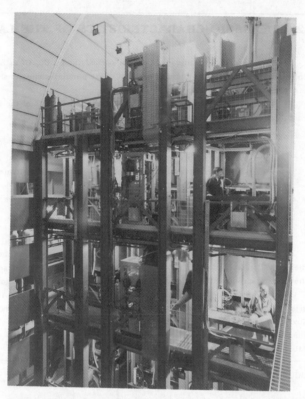

Gran Sasso Experimental Hall.

The ZEUS-Superconducting Solenoid at the HERA collider.

The ELN strategy was also focused on the activity of Italian industry which was totally absent in superconducting technology. Within a few years, Italy became prominent in this advanced field of technology with Ansaldo (Genoa) for magnet construction, Europa Metalli (Florence) for superconductors and Zanon (Vicenza) for large cryostats. This is how we could implement the construction of 250 superconducting dipoles for the HERA proton ring [10] and of the ZEUS superconducting solenoid "transparent to radiation" (the largest thin solenoid in the world) [11].

In the nuclear sector of the INFN activities, the ELN strategy allowed the Laboratori Nazionali di Legnaro (LNL) [12] in the north-east and the Laboratori Nazionali del Sud (LNS-Catania) [13] in Sicily to be equipped with the needed facilities. This activity gave rise also to the INFN-LASA laboratory in Milan [14] for R&D in superconducting technology. Let me also mention the new status (equal to that of diplomats) obtained for research physicists working abroad. These corollary results could not have been obtained without the ELN strategy.

As we will see later, the ELN strategy has produced the implementation of the LAA project, thus allowing Italy to have a project for the next supercollider (§ III).

I.2 *The old big problems now.*

Before the birth of subnuclear physics there were the big problems, called: the Fermi problem, the Rabi problem, the Landau problem, the Einstein problem and the problem of the missing mass of the universe.

During the past fifty years we have been able to build the Standard Model, i.e. the greatest synthesis of all rigorously measured and reproducible phenomena resulting from 400 years of science, since modern science began with Galileo Galilei.

On the other hand the big problems confronting us have grown bigger and bigger and their status is the following:

1) The Fermi problem: This is the gap between what is known in principle and what is discovered in practice. For example, all hadronic properties should be computable in terms of α_s of QCD and of the light quark masses. The present QCD lattice "theory" gives only generic properties of hadrons. Where are the new Fermi golden rules?

2) The Rabi problem: When the first $(g-2)$ high precision experiment proved that the muon was like a heavy electron, Rabi asked: who ordered that? Fifty years later we still have no idea. The number of such queries has even increased. Why are there three families of fermions, not less or more? Why are there families at all? Why are the gauge forces $SU(3) \times SU(2) \times U(1)$? Why do the

Professor Sidney Coleman has been awarded the Prize of "Best Lecturer" on the occasion of the 15th anniversary of the International School of Subnuclear Physics (1977).

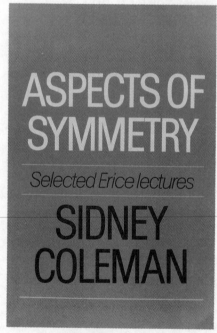

ASPECTS OF SYMMETRY

Selected Erice lectures

SIDNEY COLEMAN

CAMBRIDGE UNIVERSITY PRESS
Cambridge
New York   Port Chester
Melbourne   Sydney

Preface

I first came to Erice in 1966, to lecture at the fourth of the annual schools on subnuclear physics organized by Nino Zichichi. I was charmed by the beauty of Erice, fascinated by the thick layers of Sicilian culture and history, and terrified by the iron rule with which Nino kept students and faculty in line. In a word, I was won over, and I returned to Erice every year or two thereafter, to talk of what was past, or passing, or to come, at least insofar as it touched on subnuclear theory. Eight of these lectures, or more properly lecture series, are collected here.

No attempt has been made to bring the lectures up to date. Typographical errors, when spotted, have been corrected, and references to works to be published have been changed to references to published works. (I thank Hugh Osborne for taking on this dull task.) Otherwise, these are unaltered reprints of the original publications.

Numerous debts are acknowledged in the individual lectures, but there is one overriding debt that must be acknowledged here. None of this would have existed were it not for Nino Zichichi. Of course, he is the creator and director of the subnuclear school, and of the International Center for Scientific Culture 'Ettore Majorana' which encompasses it, but, more than that, he is personally responsible for each and every one of these lectures. The lecture notes would never have been written were it not for his blandishments and threats, transmitted in a fusillade of urgent cablegrams and transatlantic phone calls at odd hours of the morning. This book may be the least of his many accomplishments, but one of his accomplishments it is, and it should be counted as such.

Finally: These lectures span fourteen years, from 1966 to 1979. This was a great time to be a high-energy theorist, the period of the famous triumph of quantum field theory. And what a triumph it was, in the old sense of the word: a glorious victory parade, full of wonderful things brought back from far places to make the spectator gasp with awe and laugh with joy. I hope some of that awe and joy has been captured here.

Harvard 1984   Sidney Coleman

Front page of Selected Erice lectures.

basic fermionic masses range from possibly as little as a few milli-electron-Volts up to 170 GeV? Why does flavour mixing exist for quarks and leptons?

3) The Landau problem: Should we be satisfied with a theory which fails at energies inaccessibly high? Fifty years ago the Landau problem was dealing with QED. Now the problem is by far wider and refers for example to the physics which governed the earliest stages of the Big-Bang, when temperatures and densities were so great that quantum gravity had to play a basic role.

4) The Einstein problem: The Standard Model deals with only two of the three known forces. However quantum mechanics is contagious and gravity cannot avoid quantization. Lately, much of our hope has become focused on string theory. Unfortunately string theory has not yet descended to low energy, and goes on making predictions at inaccessible energies.

5) The missing Universe problem: Most of the matter in the universe is not in the form of stars, or gas, or dust. It is something else, called "dark matter", part of it is perhaps not in our present list of building blocks and glues. In a sense, there seems to be a co-existence of two "universes", independent of each other except through gravity; one with which we are familiar and the other at least 10 times heavier but of unknown matter. The present status of dark matter candidates goes from axions of mass 10^{-8} eV/c^2 to black-holes with mega-solar masses, including the three possible neutrino masses.

Unfortunately, during the last decade, in contrast with the recent past when subnuclear physics was growing at a formidable rate [see for example Erice 1977 "*The Why of Subnuclear Physics*", Plenum Press, New York and London], nothing new has been discovered: the properties of the weak bosons are exactly as expected and the number of families is exactly three. The SSC has been cancelled, without serious protests by the thousands of physicists engaged, the world over, in this unique endeavour at the frontier of human knowledge, while the LHC, a preliminary step (< 10% ELN) towards the ELN, a machine which should have by now been in operation, has had to endure attacks threatening its existence.

If we wish to avoid experiencing the end of subnuclear physics, which could unfortunately happen if we were to experience another decade such as the past one, we must have the courage to prepare the ground by planning a machine which allows us to start experimenting at (10 + 10) TeV and then keep on going to higher levels, without starting over and over again. In other words, subnuclear physics needs the maximum possible available energy spectrum to be investigated (§ III) before being abandoned.

14

The 27 km ring of LEP is superimposed on a general view of the site.

The construction at CERN of a large electron positron collider (LEP) was confronted with problems. After years of waiting, the European Committee for future Accelerators (ECFA) decided to set up an ECFA-LEP working group and appointed the author chairman, with the task of solving the problems and allowing this new European venture to be implemented. The results of the ECFA-LEP working group were published (see page 102) in what is now known as the LEP white book and this is the starting point of what was to become the largest (e^+e^-) collider in the world. One of the most significant results of LEP is the accurate measurement of the number of families:
$N_f = 2.994 \pm 0.011$ [15].

Jean Teillac (President of the CERN Council), Senator Mario Pedini (Italian Minister of Public Education) and the author (Chairman of the Working Group).

In the CERN Courier, n. 1 Vol. 19, March 1979, it is reported that the author, as «chairman of the Working Group, emphasized in his closing speech that, besides the steady scientific progress toward conceiving a wonderful tool for physics, the Meeting witnessed a strong commitment from the Italian Government to build LEP. In their interventions, the two most important Italian authorities on scientific matters — Senator Mario Pedini (Minister of Public Education) and the Hon. Dario Antoniozzi (Minister of Scientific Research) — stated that the Italian Goverment follows the LEP project very closely and hopes that its realization will not be hindered or delayed by extra-scientific matters, as has happened for other European scientific enterprises in the past.»

II – ACHIEVEMENTS IN THE PAST FIFTY YEARS.

II.1 *A TELEGRAPHIC SYNTHESIS*

Introduction.

Let me start with Table 1 where the three 1947 basic discoveries are linked to all the developments of subnuclear physics which led, fifty years later, to the Standard Model.

I was at High School when subnuclear physics, as shown in Table 1, was born in 1947. The sequence of experimental discoveries and of theoretical achievements summarized in Table 1 represents a unique challenge of mankind during these fifty years. Table 1 is the result of an unprecedented engagement of the physics community, world-wide, during these past five decades. From 1963 on, the progress in subnuclear physics has been recorded in the proceedings of the Erice schools [16]. It is thanks to the original and correct ideas which flourished during this half a century, that the roots of all Galilean (i.e. reproducible and rigorously measured) phenomena were found, in terms of three families (Fig. II.1.1) and three fundamental forces (Fig. II.1.2).

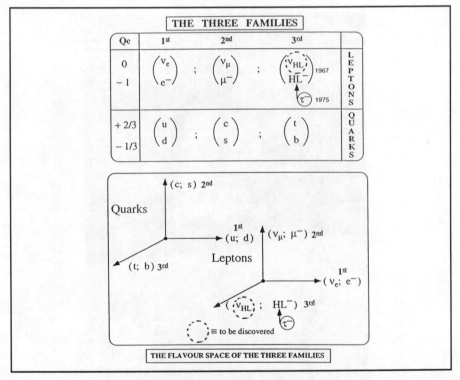

Fig. II.1.1: The Figure (from Reference 5) shows in the upper part the three families of leptons and quarks with the values of the electric charges, Qe, of each member. In the bottom part, the flavour spaces of the three families of quarks and leptons are shown. Notice that what is now called τ^- was originally defined HL^- by the BCF group [17]. Notice also that there is one missing element, never so far directly observed, the neutrino of the third family, ν_{HL}, whose existence was proposed in 1967 in the CBF proposal [17] to search for heavy leptons at Frascati.

16

Bruno Zumino at Erice, 1969.

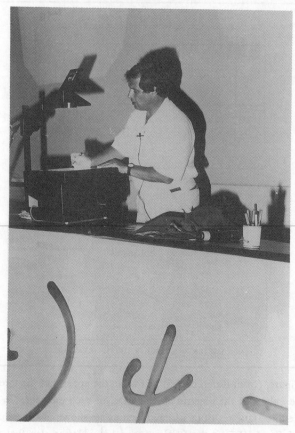

Sergio Ferrara at Erice, 1988.

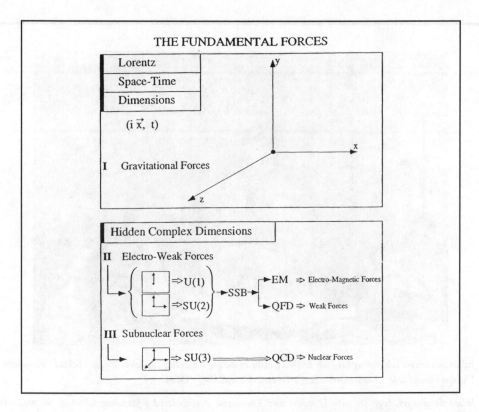

Fig. II.1.2: The Figure (from Reference 5) shows in the upper part a standard Lorentz-space-time to indicate that the gravitational forces originate from the fact that we can change at every space-time point the Lorentz $(i\vec{x}, t)$ frame with the condition that the physical results must remain the same. The origin of the other forces, the electro-weak $[SU(2) \times U(1)]$ and the strong ones $(SU(3)_c)$, is in the freedom to operate in fictitious spaces with one $(U(1))$, two $(SU(2))$ and three $(SU(3))$ complex dimensions, with the condition that physical results must remain the same. Question: what is the origin of these fictitious spaces with one, two and three complex dimensions? The story is a long one and the most recent news is in § VI Ref. 18.

All this represents the starting point of another challenge: the grand unification of all known and rigorously measured phenomena, as reported in Fig. II.1.3, where the three gauge couplings $\alpha_1\ \alpha_2\ \alpha_3$ measured with high precision at LEP (data points at M_Z) converge — if we include the radiative effects due to the superworld — towards the same origin at $E_{GUT} \simeq 10^{16}$ GeV. Here the grand unified gauge coupling turns out to have a value $\alpha_{GUT} \simeq \frac{1}{24}$. The three straight lines are the values of the gauge couplings $\alpha_1^{-1}\ \alpha_2^{-1}\ \alpha_3^{-1}$ computed on the basis of the RGEs with the EGM effect (i.e. the running gaugino masses) included, as explained in § II.2-5. The value $E_{GUT} \simeq 10^{16}$ GeV, is nearly three orders of magnitude below the Planck scale. Intense new theoretical developments are taking place in order to understand the nature of space-time at the Planck scale. These studies are linked with the physics of black holes, which have presented many mysteries in our basic notions, such as loss of information (loss of quantum coherence). This came from the discovery by Hawking (1975) [28] of what happens when a classical object (black hole)

Julian Schwinger celebrating his 70th Birthday in Erice during the 26th Subnuclear Physics School. From left: Sheldon Glashow, Mrs Manci Dirac, Sergio Ferrara, Michael Duff (1988).

Julian Schwinger, together with Feynman and Tomonaga, is considered a founding father of the modern mathematical formalism called Relativistic Quantum Field Theory (RQFT). After a period of great crisis, thanks to the discovery of the non-Abelian Gauge Forces, RQFT is now in full bloom as "effective" theory. Morever it has been discovered that the uniqueness of the S-Matrix Theory was a dream. There are as many S-Matrices as we want, all satisfying the basic principles. In fact, any non-Abelian Gauge theory, with any Gauge group and an arbitrary number of fermions (provided that they are not too many in order to avoid the loss of asymptotic freedom) will have its S-Matrix.

Freeman John Dyson during the ceremony of the Giancarlo Wick Gold Medal. Dyson, with his two papers [Phys. Rev. 75, 486, 1736 (1949)] is the physicist who was able to explain clearly the common features of the contributions of Feynman, Schwinger and Tomonaga, to the construction of RQFT (see § II.3.1).

A. ZICHICHI

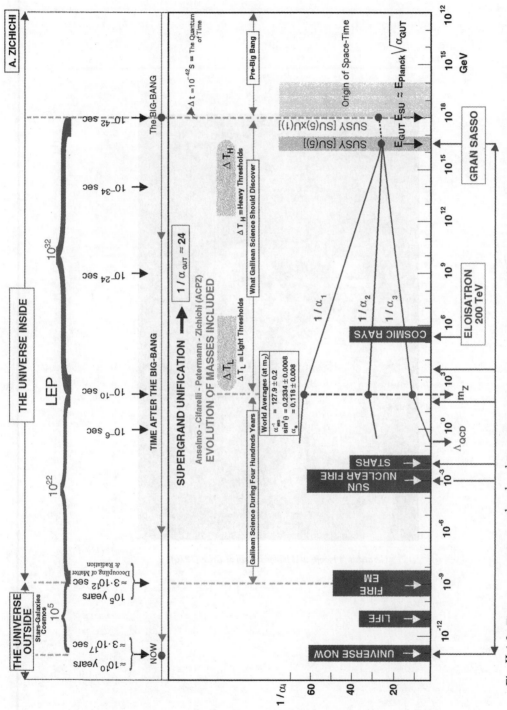

Fig. II.1.3: The gauge couplings α_1^{-1} α_2^{-1} α_3^{-1} converge towards the same origin E_{GUT}. The energy scale indicates the values corresponding to the temperature of the universe now, the energy needed for life, for the EM fire, the nuclear fire (sun), the collapse of supernovae (stars), the highest cosmic ray limits, the ELN project and the nuclear stability limit (Gran Sasso). The string unification energy (E_{SU}) is followed by the physics before the Big-Bang [29, 30] with interesting experimental consequences [31, 32].

Gerardus 't Hooft and Isidor I. Rabi at Erice (1976).

is gravitationally coupled to a quantum world, in particular to all particles of the Standard Model. String theory nowadays is shedding new light on, and thus ensuring new insight to, the quantum structure of black holes. Apparently, black holes refuse to be quantized, but G. 't Hooft is engaged in deducing space-time quantization from the very existence of a black hole [33]. Let me give another example of a theoretical development aimed at studying what could happen at the string unification scale (about 10^{18} GeV). Here, the (already unified) gauge interactions join with gravity into a single coupling α_U which, in string theory, rather than being an arbitrary parameter, is determined by the expectation value of a scalar field, the so-called dilaton $\phi(t)$ via $\alpha_U = e^{\phi(t)}$ [34]. All known string theories predict the existence of this scalar partner of the graviton. These kinds of questions have been addressed during the past few years by G. Veneziano and collaborators [29] who proposed a new revolutionary idea: the universe may have existed long before the "quantum of time" in what they call a pre-Big-Bang phase [30].

One of the most important features of this theoretical development is the prediction of observable effects, e.g. the existence of a stochastic background of gravitational waves with a characteristic frequency spectrum and interestingly high amplitudes [31], several orders of magnitude higher than that of standard inflationary cosmology.

This "gravitational light" would be the effective radiation emitted at times of the order of the Planck time ($\simeq 10^{-42}$ s). Another interesting effect is the amplification of electromagnetic fluctuations, due to the drastic variation of α_U, which could provide the long-sought explanation for the observed galactic magnetic fields [32]. These are the only ways to reach the time of the ex Big-Bang region.

In the upper part of the Figure the scale of investigation, in terms of the time distance from the old Big-Bang, is shown. Note the range which can be investigated through the studies of the universe from outside and of the universe from inside.

No matter which sophisticated instruments astrophysicists may use, they will never be able to approach the quantum of time (ex Big-Bang) below the 3×10^{12} s time distance.

At LEP, HERA and Fermi Lab, studying the universe from inside, we have reached the record of a time distance as small as 10^{-10} s from the old Big-Bang.

I am closely associated with two facilities, LEP and Gran Sasso, and to the new venture, the ELN project; this is why their role to further understand subnuclear physics is shown in the same Figure.

Table 1 and Figures II.1.1, 2, 3 are a synthesis, the explanations of which would require a very thick book, but I want to be as concise as possible to describe the total sequence of events which has brought us to the Standard Model and to think that physics will certainly go beyond it. Let me start with the "telegraphic synthesis", having as a guide Table 1 and the three Figures II.1.1, 2, 3. The stars in Table 1 (two mentioned in this chapter, the others discussed in § II) indicate some crucial developments I had the privilege of being involved in.

Remarks on Possible Noninvariance under Time Reversal and Charge Conjugation*

T. D. LEE, *Columbia University, New York, New York*

AND.

REINHARD OEHME AND C. N. YANG, *Institute for Advanced Study, Princeton, New Jersey*

(Received January 7, 1957)

4. K^0, $\bar{K}^0$ DECAY MODES

$$K_1 = \frac{1}{\sqrt{2}}(K^0 + \bar{K}^0), \quad K_2 = \frac{1}{\sqrt{2}}(K^0 - \bar{K}^0). \quad (17)$$

$$(\Gamma + iM)\psi_\pm = \lambda_\pm \psi_\pm,$$

$$\psi_\pm = \begin{pmatrix} p \\ \pm q \end{pmatrix} (|p|^2 + |q|^2)^{-1},$$

$\cdots$ One notices that *these two eigenstates ψ_+ and ψ_- do not in general represent the states K_1 and K_2 introduced in (17). In fact they even may not be orthogonal to each other.* (see footnote 11).

The above discussion also leads easily to a determination of the branching ratio of the long-lived component (and the short-lived component) into the various decay modes. $\cdots$ The branching ratio r for the decay of K_+ into $e^- + \pi^+ + \nu$ and $e^+ + \pi^- + \bar{\nu}$ is, therefore,

$$r = \frac{|pf_1 + qg_1^*|^2 + |pf_2 - qg_2^*|^2}{|pg_1 + qf_1^*|^2 + |pg_2 - qf_2^*|^2}. \quad \neq 0 \quad ,$$

---

* *Note added in proof.*—This paper was written in December, 1956, before parity nonconservation was experimentally established.

It is not very well known in the physics community that Lee, Oehme and Yang, before Parity Violation was experimentally proved by C.S. Wu, pointed out that the existence of K_2^0 could not be taken as a proof of C invariance, nor as a proof of CP invariance. Lee, Oehme and Yang (LOY) showed that "strangeness mixing" does not imply C invariance as claimed by Gell-Mann and Pais. In fact, even if CP is not valid, K_2^0 would still be there and, in order to prove that "strangeness mixing" is or is not CP invariant, other experiments needed to be done in K decay physics, as suggested by LOY. This flavour mixing problem and its CP invariance or non-invariance is extremely topical today, with many experiments being planned in order to understand the basic distinction between "flavour mixing" and CP invariance, for all flavours. Therefore we emphasize the fact that the authors of the basic distinction between "flavour mixing" and CP invariance are Lee, Oehme and Yang.

II.1-1 *We are all children of the Dirac Equation: Antiparticles and Antimatter.*

We are all children of the Dirac equation [35, 36], discovered twenty years before the birth of subnuclear physics. From that equation the first antiparticle (e^+) started to be conceived and C invariance was thought, by Dirac and Weyl [37], to be governing all physical processes. After the discovery of e^+ [38] Dirac hypothesized the existence not only of all antiparticles but also of antimatter, antistars and antigalaxies [39]. During these past seventy years since the Dirac equation, the breaking of C invariance together with P invariance (1956) [40] started a new "era" [41], where CP appeared to be conserved [42] despite that Lee, Oehme and Yang (1957) had pointed out [43] that the existence of K_2^0 could not be considered a proof of CP invariance. In fact CP invariance was discovered to be broken [44] in 1964 during the time when the apparent triumph of the S-matrix description of strong interactions [45] contributed to put in serious trouble the basic foundations of Relativistic Quantum Field Theories (RQFT); these were restricted to be of Abelian nature as the non-Abelian ones were shrouded by even more mystifying problems.

The field concept involves a larger set of functions than those derived by the analytic continuation of the S-matrix. Unfortunately no one knows how to construct fields purely in terms of analytic scattering amplitudes. Scattering amplitudes are "on the mass shell" while fields imply extension to "off the mass shell".

Let me quote a detailed example, familiar to my work: in the early sixties we determined the existence of a strong form factor of the proton in the time-like q^2 range [46, 47]. Form factors are not scattering amplitudes, nevertheless they do exist and they are due to strong interactions. The conjectured analyticity properties of the nuclear scattering matrix is a very restricted concept, if compared with the concept of a field.

S-matrix theory is not designed to describe experiments in which interactions between particle states do take place while momentum measurements are being performed. In other words all the physics due to virtual processes fell outside the physics described by the S-matrix theory.

Dirac was very concerned with the fact that, at the level of 10^{-7} and nearly a decade after the discovery of the antiproton (1956), the experimenters had observed no antideuteron production. The figure 10^{-7} is the ratio of antideuteron to π–meson production. Since no-one knew how to describe the nuclear forces in terms of a RQFT, why should these forces obey CPT invariance? In fact matter-antimatter symmetry follows from CPT invariance [48]. However this invariance is valid only if the nuclear forces can be described in terms of a Relativistic Quantum Field Theory (RQFT).

As mentioned above, the breakdown of all symmetry operators and the description of the nuclear interactions in terms of S-matrix theory (which is the negation of RQFT) gave a strong support to the possibility that nuclear antimatter was not there. As we will see later, the discovery at the 10^{-8} level of the first example of nuclear antimatter ($\bar{D}$) in 1965 [49] gave a renewed support to the validity of the CPT theorem and consequently to the

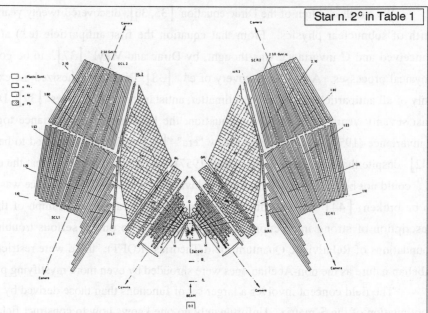

The experimental set-up implemented to search for a time-like electromagnetic structure of the proton. In this set-up the preshower for electron detection (inner part of the detector) and the muon punch-through for muon detection (lead shaped absorbers followed by heavy plate spark chambers on the external part of the set-up) are implemented on a large scale to cover a large solid angle acceptance. The set-up was able to detect simultaneously the (e^+e^-) and the $(\mu^+\mu^-)$ final states of the $(\bar{p}p)$ annihilation and, of course, the $(e\mu)$ coincidences expected from Heavy Lepton pair production (see § II.2-2).

The Split-Field-Magnet interaction region at the CERN ISR. The dE/dx counters for quarks are inside the Split-Field-Magnet.

mathematical structure RQFT for the description of the nuclear forces.

It so happens that, if in 1965 we had discovered that the $\bar{D}$ was not there, the standard model would be as it is today (see § II.2-1). This is why in Table 1 the antimatter arrow is not connected with all the other developments. The interest in the existence of antimatter has shifted from our subnuclear world to space [50]. Here it is not a question related to such fundamental problems as the invariance of the symmetry operators or the validity of RQFT. The existence of nuclear antimatter in space has to do with the universe and its evolution.

The existence of antigalaxies and antistars is far from being at the level of rigour of a theorem (like CPT); if antistars and antigalaxies exist, this depends on the cosmological description of our world and has very little impact on the standard model, which is indeed strongly linked to the three basic discoveries (1947) mentioned above. We start with the Lamb-shift.

II.1-2 *From the Lamb-shift to Scaling, Gauge Coupling Unification, the Gap and the Supersymmetry threshold.*

Let me start from the left upper part of Table 1: the Lamb-shift [1, 51]. From here follow the studies of all radiative effects which imply that the basic quantities, such as the gauge couplings $(\alpha_1\ \alpha_2\ \alpha_3)$ and the masses of quarks, leptons and gauge bosons run with energy (see § II.3-1). The discovery of scaling has played a major role in the understanding of radiative effects. In fact Deep-Inelastic-Scattering (DIS) between electrons and protons revealed a totally unexpected phenomenon. Only one "part" of the proton (called by Feynman "parton") was taking part in the interaction. The rest of the proton was totally inactive. If at high energy the proton behaves as if its pieces (partons) were "free" and therefore non-interacting among themselves, then in a high energy collision two protons should break up into their constituents, for example into "quarks".

The 1968 discovery of scaling at SLAC [52] prompted the implementation at CERN [53] of a sophisticated experimental set-up intended to establish if fractionally charged particles were "freely" produced at the highest energy (pp) collisions (using the ISR collider).

No quarks were observed by us at ISR (this contribution of my group is indicated in the Table with star nº 4, which appears twice, on the top-left in connection with *scaling*, and in the bottom right in connection with *confinement*) thus establishing a firm contradiction between the evidence that — at high energy — the pieces of the protons were losing their coupling (this is the meaning of scaling) and the evidence that no quarks were observed at ISR [53].

Meanwhile the fact that "scaling" had to be broken by radiative effects had been theoretically emphasised as if the Lamb-shift had never been discovered. Theoretical physics was in great embarrassment.

G. 't Hooft in "Under the Spell of the Gauge Principle"

«In the 1960's and early 1970's it was thought that all renormalizable field theories scale in such a way that the effective interaction strengths increase at smaller distance scales. Indeed, there existed some theorems that suggest a general law here, which were based on unitary and positivity for the Feynman propagators. I had difficulties understanding these theorems, but only later realized why. I had made some preliminary investigations concerning scaling behavior, and had taken those theories I understood best: the gauge theories. Here the scaling behavior seemed to be quite different. Now that we have the complete algebra for the scaling behavior of all renormalizable theories we know that non-Abelian gauge theories are the only renormalizable field theories that may scale in such a way that all interactions at small distances become weak. The reason why they violate the earlier mentioned theorems is that in the renormalized formulation of the propagators the ghost particles play a fundamental role and the positivity arguments are invalid. This new development was of extreme importance because it enabled us to define theories with strong interactions at large distance scales in terms of a rapidly convergent perturbative formulation at small distances.

Running of α_s

The gauge coupling α_s vs q^2 as measured in different experiments.

$$\alpha_s (M_Z) = 0.1207 \pm 0.0016 \pm 0.0066$$

[A. Wagner, Erice 1998 (to be published), from T. Doyle, ICHEP 98]

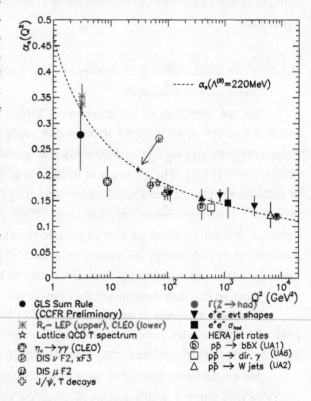

The new theory for the strong interactions based on this principle was called "quantum chromodynamics". According to this theory all hadronic subatomic particles are built from more elementary constituent particles called "quarks". These quarks are bound together by a non-Abelian gauge force, whose quanta are called "gluons". The crucial assumption was that quarks and gluons behave nearly as free particles as long as they stay close together, but attract each other with strong binding forces if they are far apart.» As a consequence, the gauge coupling α_s, must decrease, with increasing q^2.

Scaling was considered — by many theorists — to be the clear evidence that RQFT had to be abandoned. In fact it was thought that, in no case could a RQFT description of a fundamental force give rise to "scaling". Following the terminology introduced by Callan and Symanzik, a β–function with a negative sign was needed in order to have "scaling", and all RQFT were expected to produce positive β–functions. The sign of the β–function tells us if the coupling increases (positive) or decreases (negative) with k^2 (the invariant four momentum squared). If it decreases with k^2, then at $k^2 \to \infty$ the coupling vanishes (asymptotic freedom) and scaling is finally understood.

There was a young theorist, Gerardus 't Hooft, who had computed the β–function for non-Abelian theories and had found a negative value. This was in 1972, one year before Gross and Wilczek, and Politzer, announced (1973) their discovery of asymptotic freedom for a non-Abelian force able to describe the interactions between quarks and gluons; i.e. the negative sign for the β–function of $SU(3)_c$.

The most elegant way to see the sign of a β–function is the expression derived by G. 't Hooft in 1972 and presented at the Marseille conference during a discussion session:

$$\beta(g^2) = \frac{\mu d g^2}{d\mu} = \frac{g^4}{8\pi^2} \left(-\frac{11}{3} N_{colours} + \frac{2}{3} N_{fermions} + \frac{1}{6} N_{scalars} \right),$$

where the negative sign of the first term $-\frac{11}{3} N_{colours}$ is the novelty; $N_{colours}$ for SU(n) is equal to n. Thus it is 2 for SU(2) and 3 for SU(3). The other terms refer to the contributions due to the number of fermions, $N_{fermion}$, and to the number of scalars, N_{scalar}. For example, if there are no scalars and the non-Abelian gauge group is SU(2), the number of fermions can be as high as 10 and the β–function would still be negative. If the gauge group is SU(3), the number of fermions can be as high as 16 and the β–function still negative. In the notation given here, the fermions and the scalars are assumed to be in the fundamental representation[(*)] of the gauge group SU(n). This work is a turning point in our understanding

---

(*)   If there are fermions and/or scalars in the adjoint representation of SU(n) there is an extra coefficient

$$2 N_{colours}$$

in front of each term. In $SU(3)_c$ all quarks (fermions) are in the fundamental representation (triplet). Quarks could exist in the adjoint representation (octet) of $SU(3)_c$. If both types of fermions (quarks) exist, each "colour octet" is counted once with the coefficient $2 N_{colours}$:

$$\frac{2}{3} (2 N_{colours}) (N_{fermions})_{adjoint} \ ;$$

while each "colour triplet" is counted once with the coefficient $2 N_{colours} = 1$:

$$\frac{2}{3} (N_{fermions})_{fundamental}$$

where $(N_{fermions})_{fundamental}$ and $(N_{fermions})_{adjoint}$ are the number of fundamental and of adjoint representations of the fermions; the same holds for the case of scalars in the expression for $\beta(g^2)$. Another important detail: if the fermions are Majorana fermions and the scalars are not complex scalars but real scalars, then there is a factor $\frac{1}{2}$ in front of $N_{fermions}$ and of $N_{scalars}$.

"The Discovery of Scaling"
- Jerome I. Friedman in *"History of Original Ideas and Basic Discoveries in Particle Physics"*, Erice 1994, Plenum Press, 1996, p. 725.

1. EARLY ELECTRON SCATTERING EXPERIMENTS.
.... The proton resonances that we were able to measure [10] showed no unexpected kinematic behavior. Their transition form factors fell about as rapidly as the elastic proton form factor with increasing values of the four momentum transfer, q. However, we found some surprising features when we investigated the continuum region (now commonly called the deep inelastic region). ....

2. EARLY RESULTS.   2.1 Weak q^2 Dependence

The first unexpected feature of these early results [11] was that the deep inelastic cross sections showed only a weak falloff with increasing q^2. When the experiment was planned, there was no clear theoretical picture of what to expect. The observations of Robert Hofstadter and his coworkers [12] in their pioneering studies of elastic electron scattering from the proton showed that the proton had a size of about 10^{-13} cm and a smooth charge distribution. This result, plus the theoretical framework that was most widely accepted at the time, suggested to our group when the experiment was planned that the deep inelastic electron proton cross-sections would fall rapidly with increasing q^2.
....

2.2 Scaling

The second surprising feature in the data, scaling, was found by following a suggestion of James Bjorken [14]. To describe the concept of scaling, one has to introduce the general expression for the differential cross section for unpolarized electrons scattering from unpolarized nucleons, with only the scattered electrons detected [15]:

$$\frac{d^2\sigma}{d\Omega dE'} = \sigma_{Mott} \left[W_2 + 2W_1 \tan^2\tfrac{\theta}{2} \right] .$$

The functions W_1 and W_2 are called structure functions, and depend on the properties of the target system. As there are two polarization states of the virtual photon, transverse and longitudinal, two such functions are required to describe this process. In general, W_1 and W_2 are expected to be functions of both q^2 and ν, where ν is the energy loss of the scattered electron. However, on the basis of models that satisfy current algebra, Bjorken conjectured that in the limit of q^2 and ν approaching infinity, the two quantities νW_2 and W_1 become functions only of the ratio $\omega = 2M\nu/q^2$; that is

$$2MW_1 (\nu, q^2) \rightarrow F_1 (\omega) \qquad\qquad \nu W_2 (\nu, q^2) \rightarrow F_2 (\omega) .$$

The observed scaling behavior of the structure functions is shown in Fig. 2, where experimental values of νW_2 and $2MW_1$ are plotted as a function of ω for values of q^2 ranging from 2 to 20 GeV^2. The data demonstrated scaling within experimental errors for $q^2 > 2$ GeV^2 and $W > 2.6$ GeV. .... In hindsight, it is clear that these inequalities implied a point-like structure of the proton and neutron and large cross sections at high q^2, but Bjorken's results made little impression on us at the time. Perhaps it was because his results were based on current algebra, which we found highly esoteric, or perhaps it was that we were very much steeped in the physics of the time, which suggested that hadrons were extended objects with diffuse substructures.

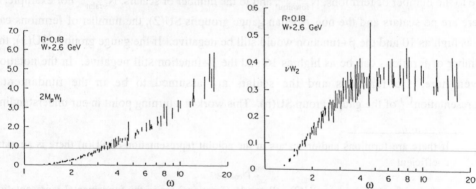

Figure 2. $2MW_1$ and νW_2 for the proton as functions of ω for $W > 2.6$ GeV, $q^2 > 1$ (GeV/c^2), and R = 0.18.

[Ref. 10: W.K.H. Panofsky, in *Proceedings of 14th Int. Conf. on High Energy Physics* Vienna (1968) 23], [Ref. 14: J.D. Bjorken, *Phys. Rev.* 179 (1969) 1547; In a private communication, Bjorken told the MIT-SLAC group about scaling in 1968].

- Conclusion: Scaling is finally understood.
Scaling is understood as a consequence of the non-Abelian nature of the gauge force acting between quarks and gluons. This is one of the most important achievements in theoretical physics, suggested by the experimental finding on the behaviour of the cross-section for inclusive processes. This behaviour was formulated by J. Bjorken in terms of the new variable $x = Q^2/2M\nu$; the physical meaning of this new variable was interpreted by R. Feynman in terms of "partons". No one could have imagined that the correct answer was in the non-Abelian nature of the forces acting inside the proton, between its constituents. For some time the "scaling" behaviour of the Deep-Inelastic-Scattering (DIS) between electrons and protons was considered by many theorists to be the first evidence for the non validity of the description, in terms of a Relativistic Quantum Field Theory, of the forces acting between the constituents of the proton. We now know that these constituents are quarks and gluons and that "scaling" is the result of the non-Abelian nature of QCD.

of RQFT; in fact, before the discovery by G. 't Hooft, the theoretical world — as mentioned above — was convinced that the sign of the β–functions of all RQFTs had to be like in QED, positive.

Scaling was finally understood as a consequence of the non-abelian nature of the force acting between the constituents of a proton (or a neutron). Consequently, the non-existence of quarks, searched for at ISR in violent collisions [53], was understood in terms of the low energy behaviour of this new force. Thus, "asymptotic freedom" and "confinement" (in § II.3-3 we will see that confinement is a consequence of imaginary mass in QCD) came in the construction of the conceptual basis of the Standard Model.

The discovery of "asymptotic freedom" awoke all physicists to the fact that the three gauge couplings $(\alpha_1\ \alpha_2\ \alpha_3)$ and not only α_3, have to run with energy. The running being different for the three couplings: α_1 increases, while α_2 and α_3 decrease when the energy increases. This gave rise to the finding that supersymmetry was needed ([54] and [55]) in order for $\alpha_1\ \alpha_2\ \alpha_3$ to converge at the same point, as it will be discussed in § II.2-5 and § II.3-1. The EGM effect, [56] i.e. the inclusion of the running of the gaugino masses, lowers the energy level where supersymmetry breaking should occur, by nearly three orders of magnitude, rendering the search for the lightest supersymmetric particle of great relevance at present energies ($\simeq 10^2$ GeV). This has to be confronted with other "predictions" — based on very primitive "geometrical" approximations [57] — claiming that the supersymmetry threshold had its best likelihood in the TeV-range.

The end point of the "Lamb-shift" arrow is in the problem of the energy gap existing between the unification scale ($\simeq 10^{16}$ GeV) of the three gauge couplings and the string unification scale ($\simeq 10^{18}$ GeV). This will be discussed in § II.2-6.

II.1-3[a] *From the π–meson to the Third Family of Leptons*.

The discovery of the π–meson appeared to be, in 1947, a triumph of field theory. The source of the nuclear force was the nucleon, the field quantum was the pion. The RQFT, originally developed to describe the electromagnetic interactions, appeared to be the natural tool for describing the dynamics of elementary particles, no matter their interactions: electromagnetic, weak or strong. After some decades of successes, RQFT started to show its severe limitations and deep troubles. The Yukawa field theory [58] was confronted with severe difficulties, such as infinities beyond the lowest order perturbation theory and the lack of any understanding of the dynamics of the nuclear forces at the non-perturbative level.

As we shall see in the chapter starting with the discovery of the V^0–particles, a totally unexpected fact came in: the rapid proliferation of strongly interacting mesons and baryons, thus depriving the nucleon and the pion fields of their privilege of being

The muon $(g-2)$ gave to the author the opportunity to meet R.P. Feynman. Zeldovich had computed that the radiative effects on the $\mu(g-2)$ due to (non-EM) couplings needed a cut-off in the GeV range if the muon had non EM interactions of reasonable strength. The muon was 200 times heavier than the electron and, at that time, the origin of the mass was unknown (as it is today) but believed to be due to extra interactions. For example, Electromagnetism was believed to produce mass differences Δm of the order of MeV, such as the (pn) mass difference. Where was the 10^2 MeV mass difference between the muon and the electron coming from? **Conclusion**: measure $(g-2)_\mu$ with good accuracy; this means something of the order of a percent. After all the muon was 200 times heavier than the electrons. There were different proposals: the screw magnet was advocated by Leon Lederman. The author felt that the "flat magnet" was the correct way out. This had already been tried by R. Nikitin in Dubna and failed because high precision in magnetic fields (at the 10^{-4} accuracy level in $\frac{\Delta B}{B}$) were needed. These fields had to be of polynomial form in order to allow injection, storage and ejection in an adiabatic way. In particular, the storage had to be a long one: the muon momentum had to rotate thousands of times, since the mis-match between linear momentum $\vec{P}_\mu$ and magnetic moment $\vec{\mu}$

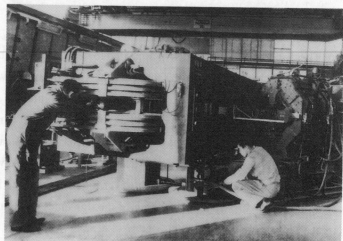

was $(\frac{\alpha}{2\pi}) \simeq 10^{-3}$ per revolution.

In order to get the 5 polynomial magnetic fields (injection, transition, storage, transition, ejection) needed (all with high precision) the experts knew one technology: high precision milling-machine to obtain the needed shape of the iron poles of a "flat magnet". The estimated time needed: many years (5-10) of very hard work. The author invented a way out, the "shimming": use very thin (microns) mu-metal sheets to achieve the "high accuracy" in the primordial shape of the "flat magnet" poles. The mu-metal was used by the author to protect the photomultipliers from the fringing magnetic fields. Dick Feynman was a theorist very interested in experimental tricks. He was very much pleased with the idea of constructing high precision polynomial magnetic fields using the trick which avoided very expensive and time-consuming highly accurate machining. Once Nikitin came to CERN and visited the $(g-2)$ flat magnet (not the 6 metres long one, but the "Liverpool" magnet, about one metre long, graciously lent by Liverpool University to CERN in order to allow our tests). He told the author that he had tried to use a flat magnet for magnetic "storage" and failed. Thus giving a friendly advice to stop wasting time. Lederman was on his side and in fact used this advice to start his "screw magnet" technology. Later, Feynman in Weisskopf's office was much more encouraging: «never follow the advice of people who have failed in what you want to do».

This is how the author and Feynman became friends: Feynman came to the Ettore Majorana School during the hard times when the Ettore Majorana School was holding its second course (1964).

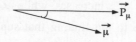

Star n. 2ª in Table 1

A photo of the six-metre "flat-magnet" where a sequence of high precision magnetic fields have been implemented using the "trick" of the "shimming technology".

"fundamental".

The formidable role of the π–meson turned out to be in the field of lepton physics. The π discovery provided the "glue" for the nuclear forces and this was great. This discovery was followed by the observation of the complete decay-chain-reaction

$$\pi \rightarrow \mu \rightarrow e \, ,$$

which allowed one to understand the real nature of the cosmic ray "meson" discovered in 1936 by Anderson and Neddermeyer [59]. Despite the war, Conversi, Pancini and Piccioni were able to perform in Italy an experiment where the negative component of the cosmic ray "meson" was stopped and observed to decay as if it had no nuclear coupling with matter [60]. This cosmic ray "meson" was not Yukawa's meson but its decay product, as finally proved by the complete decay-chain-reaction above. It was G. Puppi who first pointed out that the Fermi couplings of different weak processes, including the muon, were the same [61], within an order of magnitude. The identification of the muon as a particle deprived of the strong force opened the problem of the second lepton "μ". On the other hand, if it were not for the π–meson, no-one would have known the existence of this second lepton. As we shall see later, the accurate measurements of its electromagnetic [$(g-2)_\mu$] [66-70] and weak (τ_μ) [71, 72] properties are at the origin of the idea that an intensive search for a heavier lepton, not having the privilege of being produced by another "ad-hoc" meson, had to be implemented. We shall discuss this in § II.2-2. It is probably interesting to mention that the first high precision check of the electromagnetic properties of the muon was performed via the measurement of its anomalous magnetic moment. This experiment required the construction of the largest and highest precision "flat" magnet of the world, whose schematic drawing is reported in Fig. II.1-3^a.1.

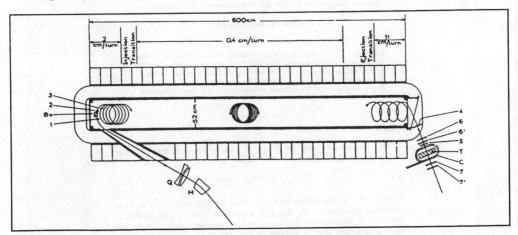

Fig. II.1-3^a.1: (Figure from Reference 67). General plan of the 6-metre magnet. M: bending magnet; Q: pair of quadrupoles; 1, Be, 2, 3: injection assembly consisting of Be-moderator and counters 1, 2, 3; T: methylene-iodide target; counters 66', 77': "backward" and "forward" electron telescopes. A stored and ejected muon is registered as a coincidence 4, 5, 66'$\overline{7}$, gated by a 1, 2, 3 and by either a forward or backward electron signal.

John Stewart Bell at Erice (1963).

IL NUOVO CIMENTO   VOL. LX A, N. 1   1º Marzo 1969

A PCAC Puzzle: $\pi^0 \rightarrow \gamma\gamma$ in the σ-Model.

J. S. Bell

CERN - Geneva

R. Jackiw (*)

CERN - Geneva

Jefferson Laboratory of Physics, Harvard University - Cambridge, Mass.

(ricevuto l'11 Settembre 1968)

Summary. — The effective coupling constant for $\pi^0 \rightarrow \gamma\gamma$ should vanish for zero pion mass in theories with PCAC and gauge invariance. It does not so vanish in an explicit perturbation calculation in the σ-model. The resolution of the puzzle is effected by a modification of Pauli-Villars-Gupta regularization which respects both PCAC and gauge invariance.

---

1. – Introduction.

The invariant amplitude for $\pi^0 \rightarrow \gamma\gamma$ is obtained by contracting the polarization vectors of the photons with a tensor

$$(1.1) \qquad T^{\mu\nu}(p, q) = \epsilon^{\mu\nu\alpha\beta} p_\alpha q_\beta T(k^2),$$

where p and q are the photon momenta, which we shall always take to be on their mass shell $p^2 = 0 = q^2$. The pion momentum is $k(= p + q)$; we shall be interested in off-mass-shell values, as well as physical values $k^2 = \mu^2$. The above general form of $T^{\mu\nu}$ is dictated by Lorentz invariance and parity conservation. Gauge invariance ($p_\mu T^{\mu\nu} = T^{\mu\nu} q_\nu = 0$) and Bose symmetry ($T^{\mu\nu}(p, q) = T^{\nu\mu}(q, p)$) are seen to hold.

(*) Junior Fellow, Society of Fellows.

II.1-3^b *From the π^0 to the ABJ Anomaly and to the Instantons.*

To explain the experimental observation of the charge independence of nuclear forces, Kemmer pointed out [73] that, in addition to Yukawa's charged meson, a neutral pion had to exist. The production of a neutral meson decaying into two photons was suggested by Lewis, Oppenheimer and Wouthuysen [74] in order to explain the soft component in cosmic radiation. The missing member of the pion triplet was discovered in cosmic rays at Bristol by Carlson (Ekspong since 1951) et al. [75] and, using the cyclotron at Berkeley, by Bjorklund et al. [76], and by Panofsky et al. [77]. *«It was generally felt that the neutral pion marked the end for particle searches»* [78]: this is what was said, forty-seven years after its discovery, by a prominent member of the physicists who discovered the neutral pion.

Instead of the end of particle searches, the $\pi^0 \to \gamma\gamma$ marked the opening of a new horizon in subnuclear physics. The π^0 decaying into 2γ inspired Julian Schwinger to think: the π^0 was a pseudoscalar and the two photons were linked to an electromagnetic current. One year after the discovery of the $\pi^0 \to \gamma\gamma$ Julian Schwinger [79] proved that, if in QED the current was axial, the conservation of the axial charge (an immediate consequence of axial symmetry) would not give conservation of the axial electromagnetic current (when the current operator is appropriately regularized).

In 1963 Ken Johnson proved that in 2-dimensional QED, either you conserve the vector current (the gauge current) or the axial current, not both (as we will see later something similar happens in 4 dimensions).

In the sixties, Gell-Mann and Lévy [80] invented the σ model to explain the current algebra results, the most important being PCAC (Partial Conservation of the Axial Current); independently, the same PCAC hypothesis was also proposed by Chou Kuang-Chao [81]. Sutherland [82] and Veltman [83] proved the theorem that if the σ model was a valid approach to physics, the neutral pion, discovered in 1950 to decay into 2γ, cannot decay. In 1969 J.S. Bell and R. Jackiw [84] found that the σ model, interpreted in the conventional way, does not satisfy PCAC. In the current algebra calculations it was assumed that it was possible to have the conservation of the three currents (one axial and two vectorial) and this is the origin of the Sutherland and Veltman theorem. Bell and Jackiw discovered the quantum mechanical breaking of the axial symmetry — i.e. the by now famous anomaly — which allows $\pi^0 \to \gamma\gamma$ to exist. In the same year, Stephen Adler, working in spinor electrodynamics, also discovered the anomaly [85]. The crucial point is that, if you compute the triangle Feynman diagram (Fig. II.1-3^b.1) made up of one axial and two vector currents containing an UV divergence, it happens that, while the conservation of the vector current can be maintained, the conservation law of the axial current is broken: something very similar to the Johnson result quoted above.

PHYSICAL REVIEW   VOLUME 177, NUMBER 5   25 JANUARY 1969

Axial-Vector Vertex in Spinor Electrodynamics

STEPHEN L. ADLER

Institute for Advanced Study, Princeton, New Jersey 08540

(Received 24 September 1968)

Working within the framework of perturbation theory, we show that the axial-vector vertex in spinor electrodynamics has anomalous properties which disagree with those found by the formal manipulation of field equations. Specifically, because of the presence of closed-loop "triangle diagrams," the divergence of axial-vector current is not the usual expression calculated from the field equations, and the axial-vector current does not satisfy the usual Ward identity. One consequence is that, even after the external-line wave-function renormalizations are made, the axial-vector vertex is still divergent in fourth- (and higher-) order perturbation theory. A corollary is that the radiative corrections to ν_l elastic scattering in the local current-current theory diverge in fourth (and higher) order. A second consequence is that, in massless electrodynamics, despite the fact that the theory is invariant under γ_5 transformations, the axial-vector current is not conserved. In an Appendix we demonstrate the uniqueness of the triangle diagrams, and discuss a possible connection between our results and the $\pi^0 \rightarrow 2\gamma$ and $\eta \rightarrow 2\gamma$ decays. In particular, we argue that as a result of triangle diagrams, the equations expressing partial conservation of axial-vector current (PCAC) for the neutral members of the axial-vector-current octet must be modified in a well-defined manner, which completely alters the PCAC predictions for the π^0 and the η two-photon decays.

William A. Bardeen.

PHYSICAL REVIEW   VOLUME 184, NUMBER 5   25 AUGUST 1969

Anomalous Ward Identities in Spinor Field Theories

WILLIAM A. BARDEEN*†

Institute for Advanced Study, Princeton, New Jersey 08540

(Received 24 February 1969)

We consider the model of a spinor field with arbitrary internal degrees of freedom having arbitrary nonderivative coupling to external scalar, pseudoscalar, vector, and axial-vector fields. By carefully defining the S matrix in the interaction picture, the vector and axial-vector currents associated with the external vector and axial-vector fields are found to satisfy anomalous Ward identities. If we require that the vector currents satisfy the usual Ward identities, the divergence of the axial-vector current contains well-defined anomalous terms. These terms are explicitly calculated.

I. INTRODUCTION

THE presence of anomalous terms in the Ward identities for currents defined in a number of spinor field theories has been noted by several authors.[1-3] The existence of these terms may be traced to the local products of field operators which are so singular as to prohibit the naive use of the field equations. In a version of the σ model, the anomalous terms in the Ward identity for the neutral isospin current have led to a low-energy theorem for the decay $\pi^0 \rightarrow \gamma\gamma$.[2]

In this paper, we consider a theory of a spinor field with an arbitrary number of internal degrees of freedom coupled to external scalar, pseudoscalar, vector, and

* Research sponsored by the Air Force Office of Scientific Research, U. S. Air Force, under AFOSR Grant No. 68-1365.

† Present address: Department of Physics, Stanford University, Stanford, California 94305.

[1] J. Steinberger, Phys. Rev. 76, 1180 (1949); J. Schwinger, *ibid.* 82, 664 (1951).

[2] S. L. Adler, Phys. Rev. 177, 2426 (1969).

[3] C. R. Hagen, Phys. Rev. 176, 2622 (1969); R. Jackiw and K. Johnson, *ibid.* 182, 1457 (1969); R. Brandt, *ibid.* 180, 1490 (1969); K. Wilson, *ibid.* 181, 1909 (1969); J. S. Bell and R. Jackiw, Nuovo Cimento 60, 47 (1969).

The ABJ anomaly opened the door to a deeper understanding of RQFT: a new "era" for RQFT started. An interesting property of the anomaly was proved soon afterwards by S. Adler and W.A. Bardeen: the anomaly receives no contributions from radiative corrections of any order; i.e. the anomaly is totally given by the 1-loop calculation shown in Fig. II.1-3^b.1 [86].

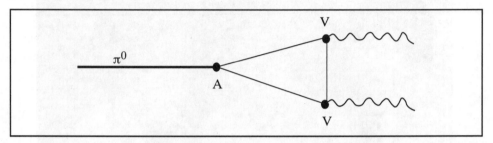

Fig. II.1-3^b.1: The full description of the anomaly is given by the triangle graph where there are 3 vertices, one axial (A) and two vectorial (V). No higher order radiative diagrams contribute to the chiral anomaly.

This is an important result because it is valid not only in QED, but also in QCD. For more general theories the Adler-Bardeen theorem needs to be confirmed, but insofar as we deal with the Standard Model the result remains valid.

The extension of the "chiral anomaly" — this extraordinary new horizon opened by the discovery of the $\pi^0 \to \gamma\gamma$ — to non-Abelian fields was started by W.A. Bardeen [87] and is an important ingredient in model building. For example, in the Standard Model it enforces colour triality and explains why the number of fundamental quark-fermions must be equal to the number of fundamental lepton-fermions, thus allowing the prediction of the last quark, in addition to the "b" quark in the 3rd family: i.e. the top-quark.

The ABJ anomaly, revealed by the existence of $\pi^0 \to \gamma\gamma$, has produced an impressive set of consequences in physics and mathematics. The most interesting is the existence of new "topological" solutions. Algebraic topology is a very rich subject. It provides stable solutions in one, two, three and four dimensions. Stable objects, in one dimension, correspond to "domain walls". In two dimensions, to "vortices". In three dimensions, to "magnetic monopoles" [88, 89]. In four dimensions, the stable solutions are the "instantons" [90, 91] which provide "tunnelling transitions" between energy levels in the Dirac sea. These transitions produce the violation of conservation laws. For the weak interactions this means violation of baryon number and lepton number conservation [92]. In QCD the "instantons" induce the direct (non-spontaneous) violation of chiral charge. This allows to understand why the π is not a perfect Goldstone boson (i.e. massless) and the η and the η' are so heavy [93]. As mentioned above, the Instantons originate from a deep analysis of what really is the Dirac vacuum: i.e. the lowest energy state made up

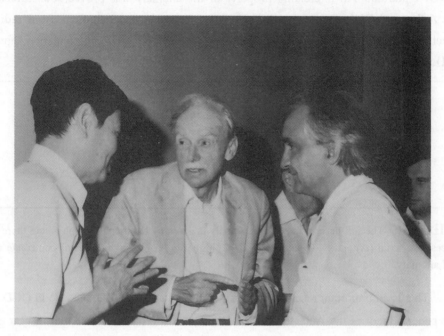

Chen Ning Yang with Paul Adrien Maurice Dirac at Erice (1982).

Val Fitch at Erice (1981).

James W. Cronin.

of fermions. We will return later to this topic in § II.2-4 and § II.3-5. To sum up, without the discovery of the $\pi^0 \rightarrow \gamma\gamma$ this goldmine of subnuclear physics could not have been found.

II.1-4[a] *From the V^0–particles to C breaking, P breaking, the K–meson complex, with Flavour Mixing, CP violation and T breaking.*

The V^0–particles gave rise to the $(\theta-\tau)$ problem [94], which culminated in the discovery of the breaking $(\neq)$ of the symmetry operators C and P. The discovery of the non-invariance of these symmetry operators was suggested (1956) in a detailed analysis of all weak processes by T.D. Lee and C.N. Yang [40]; the first experimental evidence was provided by C.S. Wu and collaborators one year later [41]. Nevertheless, a lot of confusion was around in those years. For example the "strangeness mixing", proposed by Gell-Mann and Pais [95], to describe the $(\theta^0\overline{\theta}^0)$ pair, brought them to predict the existence of θ_1^0 and θ_2^0, on the basis of the validity of C invariance in weak interactions. The discovery by Lederman of $\theta_2^0 \rightarrow 3\pi$ [96] was interpreted as a proof that C invariance holds in weak interactions. With the discovery of C and P breaking, the $(\theta-\tau)$ mesons became a unique particle, the K–meson, which split into two components, K_1^0 and K_2^0, each one thought to be an eigenstate of the symmetry operator CP proposed by Landau [42] to replace the two broken P and C invariances. Probably few people know that, in 1956, Lee, Oehme and Yang, before parity violation was experimentally proved by C.S. Wu, pointed out that the existence of K_2^0 could not be taken as a proof of C invariance, nor as a proof of CP invariance [43]. Lee, Oehme and Yang (LOY) showed that "strangeness mixing" does not imply C invariance as claimed by Gell-Mann and Pais. In fact, even if CP is not valid, K_2^0 would still be there and, in order to prove that "strangeness mixing" is or is not CP invariant, other experiments had to be done in K decay physics, as suggested by LOY. In 1964, it was discovered that CP and T invariances are broken [44].

Let me quote an amusing detail of this great discovery. The experiment was not planned to search for the 2π decay mode of the K_2^0 meson. The aim of the experiment was to check the anomalous regeneration in hydrogen, previously reported by Robert Adair et al. [97] (and found [44] to be more than an order of magnitude lower). The search for the 2π decay mode of the long-lived K_2^0 was proposed at CERN, but rejected because the neutral beam in the PS experimental hall had already been allocated to another group's programme. On the other hand, we were already engaged with the PAPLEP (Proton AntiProton Annihilation into LEpton Pairs) experiment to search for the production of the 3rd lepton through the $(e\mu)$ final state produced in $(\overline{p}p)$ annihilation [98], using the CERN-PS beam which was next to the neutral beam we wanted for the $K_2^0 \rightarrow 2\pi$ search. I was told by the

Murray Gell-Mann at Erice (1973).

From left: Luigi Dadda, Pierre A. Piroué, Enrico Bignami, Yuval Ne'eman, Richard L. Garwin, John C. Eccles, Eugene P. Wigner, the author, Edward Teller, Paul Adrien Maurice Dirac, George Charpak (1981).

CERN Research Director of the time "give other people the chance", when trying to convince him that the existence of the long lived K_2^0 was not proof of CP invariance as shown by LOY in 1957 [43], therefore the search for the $K_2^0 \rightarrow 2\pi$ decay mode violating CP invariance was not in contradiction with the existence of the long lived K_2^0 meson. It would have been too much to give two PS beams to the same group, he told me later. Moreover we were not proposing to check the anomalous regeneration in hydrogen (a proposal considered very interesting). Our aim was to follow the LOY theoretical deep remark and check if CP was really valid.

The flavour mixing problem and its CP invariance or non-invariance, is extremely topical today with many experiments being planned in order to understand the basic distinction between "flavour mixing" and CP invariance, for all flavours. The fact that the authors of the basic distinction between "flavour mixing" and CP invariance are LOY has been forgotten. How and why the quark flavours (u, c, t) and (d, s, b) mix and why this mixing is linked with the breaking of CP has no theoretical understanding, so far. All we can do is measure the various flavour mixings and the CP breakings.

As we shall see in § II.3-4, flavour mixing appears to be active also in the lepton sector. Sooner or later, these problems need to be understood.

II.1-4^b *From the V^0–particles to $SU(3)_f$ and $SU(3)_c$.*

Another chain of consequences originated by the existence of the V^0–particles was the proliferation of mesons and baryons with two branches: "statics" and "dynamics".

The "static" proliferation gave rise, first to an order of magnitude reduction of the mesonic and baryonic states via the eightfold way of Gell-Mann and Ne'eman [99], and then to the "flavour" global symmetry $SU(3)_f$ based on the existence of three quark flavours: u, d, s [101].

We know that the number of quark flavours is six: u d c s t b, thus the "static" reduction of proliferation via $SU(3)_{f\ =uds}$ was an illusion.

Nevertheless $SU(3)_f$ contributed to open the way towards $SU(3)_c$. It is in fact the notion that two baryons

$$\Omega^- \quad \text{and} \quad \Delta_{3/2\,3/2}^{++}$$

had to be fermions, but appeared to be perfectly symmetric in their quark composition [102], that prompted the idea for the existence of a new intrinsic quantum number [103, 104].

This V^0–chain of consequences, linked with the other (scaling) in the Lamb-shift chain, led to the discovery [105] of $SU(3)_c$, whose manifestation is quantum

- 1225 -

GAUGE THEORIES WITH UNIFIED WEAK, ELECTROMAGNETIC
AND STRONG INTERACTIONS

G. 't Hooft
Institute for Theoretical Physics
University of Utrecht
Utrecht, The Netherlands

1. INTRODUCTION

Only half a decade ago, quantum field theory was considered as just one of the many different approaches to particle physics, and there were many reasons not to take it too seriously. In the first place the only possible "elementary" particles were spin zero bosons, spin $\frac{1}{2}$ fermions, and photons. All other particles, in particular the ρ , the N^X , and a possible intermediate vector boson, had to be composite. To make such particles we need strong couplings, and that would lead us immediately outside the region where renormalized perturbation series make sense. And if we wanted to mimic the observed weak interactions using scalar fields, then we would need an improbable type of conspiracy between the coupling constants to get the V-A structure[1]. Finally, it seemed to be impossible to reproduce the observed simple behaviour of certain inclusive electron-scattering cross sections under scaling of the momenta involved, in terms of any of the existing renormalizable theories[2]. No wonder that people looked for different tools, like current algebra's, bootstrap theories and other nonperturbative approaches.

Theories with a non-Abelian, local gauge invariance, were known[3], and even considered interesting and suggestive as possible theories for weak interactions[4,5], but they made a very slow start in particle physics, because it seemed that they did not solve very much since unitarity and/or renormalizability were not understood and it remained impossible to do better than lowest order calculations.

When finally the Feynman rules for gauge theories were settled[6] and the renormalization procedure in the presence of spontaneous symmetry breakdown understood[7-19] it was immediately realized that there might exist a simple Gauge Model for all particles and all interactions in the world. The first who would find The Model would obtain a theory for all particles, and immortality . Thus the Great Model Rush began[29,35,36,44-53,56,58].

First, one looks at the leptons. The observed ones can easily be arranged in a symmetry pattern consistent with experiment[5]: SU(2) x U(1). But if we assume that other leptons exist which are so heavy that they have not yet been observed then there are many other possibilities. To settle the matter we have to look at the hadrons.

The observed hadron spectrum is so complicated with its octuplets, nonets and decuplets that it would have been a miracle if they would fit in a simple gauge theory like the leptons. They don't. To reproduce the nice

Front page of the EPS Proceedings. These proceedings are the first and exhaustive record of the status of Subnuclear Physics, immediately following the "November revolution". For example G. 't Hooft in his report starting on page 1225 discusses for the first time the introduction of imaginary masses in QCD to explain "confinement".

chromodynamics (QCD) [106].

In this great construction of QCD there was a "hidden" side and this refers to the proliferation in the "dynamics": how can it be that a unique fundamental force acting among its two very simple basic components, quarks and gluons, produces such a variety of final states as those observed when a pair of particles interact?

It is the effective energy which allows to overcome this "hidden" trouble of QCD, as we shall see in § II.2-3.

We have just mentioned in § II.1-2 the final theoretical trouble affecting QCD: confinement. From QCD it is not possible to predict confinement. But our ISR results [53] prove that, at high energy, confinement holds. So, QCD must produce: i) Asymptotic freedom, in order to explain SLAC scaling, and ii) Confinement, in order to explain the ISR results of no quarks at high energy. It was in 1975, at the EPS Palermo conference, that G. 't Hooft presented his way out to explain confinement [107]. Assume that, in QCD, in addition to the quarks, there are scalar particles with imaginary masses and "colour-magnetic" charges. If this happens the colour-magnetic-QCD charges, monopoles, condense, thus providing permanent confinement for the QCD-colour-electric charges, i.e. the quarks.

II.2 _TOPICS WELL KNOWN TO ME_

Introduction.

This section is easy for me since it deals with topics I had the privilege to be directly involved in. The first one is nuclear antimatter, where the construction of the most intense beam of negative particles at CERN was needed. It is followed by the problem of the third lepton to which I have devoted ten years of my scientific life. The effective energy refers to what Gribov called the "hidden" side of QCD. The gauge coupling unification $(\alpha_1 \, \alpha_2 \, \alpha_3)$ and the gap problem refers to a "hobby" of mine, related to the energy threshold for supersymmetry.

II.2-1 *Nuclear Antimatter (1965).*   ┌─────────────────────┐
                                        │ Star n. 1 in Table 1 │
                                        └─────────────────────┘

In the sixties the elementary particles were objects such as the proton (p) and the neutron (n), with antiprotons $(\bar{p})$ and antineutrons $(\bar{n})$ as antiparticles. The existence of nuclear matter[(*)] needs a nuclear binding between protons and neutrons. The mass

---

[(*)]   The hydrogen atom needs the masses of two elementary particles, the proton (p) and the electron (e), plus the electromagnetic binding between them (p, e). The existence of the hydrogen atom has nothing to do with the existence of nuclear matter.

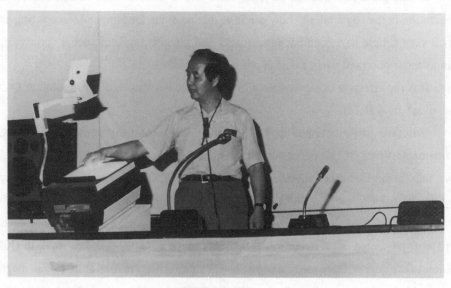

Tsung Dao Lee.

Conference Proceedings Vol. 53
-The Discovery of Nuclear Antimatter-
L. Maiani and R.A. Ricci (Eds.)
SIF, Bologna, 1996

ARE MATTER AND ANTIMATTER SYMMETRIC?

Tsung Dao Lee

Columbia University, New York, NY 10027

President Barnabei, Rector Roversi Monaco and distinguished guests.

It is a great pleasure for me to be here to participate in this symposium in honor of my good friend, Professor Antonino Zichichi, and to celebrate the 30th anniversary of his discovery of nuclear antimatter.

We note that particles and antiparticles refer to the elementary units of which all matter and antimatter consist. Thus, the symmetry between the latter implies that between the former, but not vice versa. Consequently, it is of fundamental importance to test the symmetry between matter and antimatter, as forcefully expressed in the following quotation from Heisenberg (in *The Physicist's Conception of Nature* (1972)):

«I think that this discovery of antimatter was perhaps the biggest jump of all the big jumps in physics in our century.»

Now, Heisenberg discovered quantum mechanics in 1925. By 1972, he had witnessed almost all the big jumps in modern physics. Yet he would rank the discovery of antimatter as the biggest jump of all.

The demarcation between particles and matter changes with time. The history of the question of symmetry between matter and antimatter is bound to be interwoven with that between particles and antiparticles. Thus, we have to begin with the landmark work

1

of the most elementary nucleus of matter, the deuteron (D), needs, in addition to the masses of the two elementary particles (p, n), also the negative nuclear mass produced by their binding. As mentioned in § II.1-1, in the sixties, there was no understanding of the mathematical structure needed to describe these nuclear binding forces.

Since the middle sixties, our understanding of the nuclear binding forces has evolved a lot, as we have just seen, thanks to QCD. And now a problem arises. The basic ingredients of QCD are the gluons (massless) and the quarks (massive). No one knows the scale where the intrinsic quark masses originate. If it is at the string unification scale (i.e. $\simeq 10^{19}$ GeV), the CPT theorem loses its foundations (T.D. Lee, 1995 [108]) and therefore, although particle theories with CPT violation are not presently known, it is yet of importance to check the equality of masses for particles and antiparticles. In fact, it could be that,

$$m_q \neq m_{\bar{q}} \, .$$

Let us disentangle:

i)   the intrinsic mass associated with a quark (a structureless particle);

from

ii)   the mass associated with a nucleon   (a particle composed of three quarks plus many gluons);

and these two masses from

iii)   the mass associated with nuclear matter, the simplest example being the deuteron.

It is sometimes stated that the existence of a mass difference between the long-lived and the short-lived components of the $(K^0 \bar{K}^0)$ system is the proof that matter-antimatter symmetry is broken. The experimental result is:

$$\Delta m_{K_L K_S} = m_{K_L} - m_{K_S} = (3.491 \pm 0.009) \times 10^{-6} \, eV/c^2 \, . \tag{1a}$$

However this is the mass difference between two particle states, K_L and K_S, each one consisting of a mixture of a particle (K^0) and its antiparticle ($\bar{K}^0$). When (1a) is translated into the mass difference between the K^0 and the $\bar{K}^0$ the result is:

$$\Delta m_{K\bar{K}} = \left| m_{K^0} - m_{\bar{K}^0} \right| \lesssim 4 \times 10^{-10} \, eV/c^2 \, . \tag{1b}$$

In other words there is no final statement (in the case of a meson) for the existence of any asymmetry between the mass of a particle (K^0, i.e. a $q_i \bar{q}_j$ system) and its antiparticle ($\bar{K}^0$, i.e. a $\bar{q}_i q_j$ system). Let us point out again that, what in the middle sixties was considered an elementary particle, is now understood to be a system of either a quark-antiquark ($q\bar{q}$) pair (mesonic state) bound by QCD colour confining forces, or a (qqq) triplet (baryonic state) bound by QCD colour confining forces. The masses of these particles (mesons and baryons) are the result of the intrinsic quark masses, m_q, plus the QCD confining ("Bag") effects, m^{Bag}, plus some radiative effects, m^{Rad}.

Oreste Piccioni and the author at the 1956 International Conference in Geneva.

Paul Adrien Maurice Dirac in Erice (1982).

As a meson is already a mixture of a quark plus an antiquark, the search for an asymmetry between particle and antiparticle masses should have its best source in those particle states which consist only of quarks (such as the baryons), and not of quark-antiquark mixtures (such as the mesons) [109].

Keeping in mind the problem of the deuteron and antideuteron masses, let us consider the mass difference between a particle and its antiparticle, each one composed of quarks and gluons. The simplest example is the proton, whose mass is the result of the following components:

$$m_p \equiv 2\, m_u + m_d + m_{uud}^{Bag} + m_{uud}^{Rad} \qquad (2a)$$

where:

i) m_u, m_d are the intrinsic masses of the elementary constituents, the quarks;

ii) m_{uud}^{Bag} is the mass produced by the QCD colour forces acting between quarks and gluons and confining them within the proton radius;

iii) m_{uud}^{Rad}, has been defined earlier.

The same parts appear in the mass of an antiproton:

$$m_{\bar{p}} \equiv 2\, m_{\bar{u}} + m_{\bar{d}} + m_{\bar{u}\bar{u}\bar{d}}^{Bag} + m_{\bar{u}\bar{u}\bar{d}}^{Rad} \ . \qquad (2b)$$

If the interaction responsible for the intrinsic mass of a quark is CPT invariant, if the QCD confining effects and the radiative effects are all CPT invariant, the result is expected to be

$$\Delta m_{p\bar{p}} = m_p - m_{\bar{p}} = zero \ ;$$

the experimental limit is:

$$\Delta m_{p\bar{p}} = (2.2 \pm 40)\ eV/c^2 \simeq zero \pm 40\ eV/c^2 \ . \qquad (2c)$$

And now, the deuteron-antideuteron masses:

$$m(D) = m_p + m_n - m_{pn}^{Binding} + m_{pn}^{Rad} \qquad (3a)$$

$$m(\bar{D}) = m_{\bar{p}} + m_{\bar{n}} - m_{\bar{p}\bar{n}}^{Binding} + m_{\bar{p}\bar{n}}^{Rad} \ . \qquad (3b)$$

In addition to the particle masses (m_p, m_n) and the antiparticle masses $(m_{\bar{p}}, m_{\bar{n}})$, we now have the nuclear binding effects, $m_{pn}^{Binding}$ and $m_{\bar{p}\bar{n}}^{Binding}$, which, contrary to the QCD "bag" effects (that produce positive masses), subtract mass from the $(p\, n)$ and $(\bar{p}\, \bar{n})$ systems, respectively.

If all these processes are CPT invariant, we expect the mass difference between the deuteron and the antideuteron to be zero

$$\Delta m_{D\bar{D}} = m_D - m_{\bar{D}} = zero \ .$$

The experimental limit is:

$$\Delta m_{D\bar{D}} = zero \pm 80\ MeV/c^2 \ . \qquad (3c)$$

46

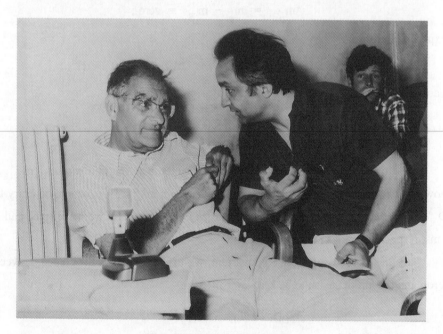

Victor F. Weisskopf in Erice (1963).

It would be interesting in future to see how these results [(1b); (2c); (3c)] compare among themselves, once they have reached the needed sensitivity. Notice that the mass uncertainty in (1a) is ($\pm 9 \times 10^{-9}$ eV), i.e. nearly ten orders of magnitude lower than the value (± 40 eV) which characterizes the best mass difference so far measured in a particle-antiparticle system made up of three quarks (2c). Apart from being — as already emphasized — a mass difference between two particle states (not between a particle and its antiparticle), the reason for the extraordinary accuracy in $\Delta m_{K_L K_S}$ is in the fact that what is measured is a time-dependent "oscillation", whose value depends on Δm. Nevertheless, neither (1a, b) nor (2c) are measurements of mass differences between nuclear matter and nuclear antimatter states. In fact, the QCD-induced nuclear binding, which produces effects opposite in sign to the QCD confining forces, is absent in (1a, b) and (2a, b, c).

To recapitulate, in these last thirty years, our understanding of the mass differences between particle-antiparticle and matter-antimatter states has developed and, apart from radiative effects, can be described in terms of three sources:

  i)   the intrinsic mass of some fundamental fermions (the quarks): m_q ;
  ii) the "bag" effects due to QCD confining colour forces; these effects produce positive masses: m_{BAG} ;
  iii) the binding effects due to QCD colour-neutral states (the mesons) acting between other QCD colour-neutral states (the nucleons); these effects produce negative masses: $m_{Binding}$ (the negative sign is explicitly shown in Fig. II.2-1.1).

All these sources of masses, the intrinsic fermionic ones, the QCD colour confining ones and the QCD colour-neutral binding, appear in nuclear matter and antimatter, as illustrated in Fig. II.2-1.1.

$$
\begin{array}{ccccccc}
m_q & + & m_{BAG} & - & m_{Binding} & = & m_{MATTER} \\
m_{\bar{q}} & + & m_{\overline{BAG}} & - & m_{\overline{Binding}} & = & m_{\overline{MATTER}}
\end{array}
$$

Fig. II.2-1.1: The three components in matter and antimatter masses.

Of these three sources of masses, the one which produces the intrinsic quark masses $m_q \equiv m_{Intrinsic}$ is unknown and certainly not due to QCD. The other two have the same origin, QCD, but are generated by drastically different QCD effects. It is clear that

$$ m_{Intrinsic} \neq m_{Confining} \neq m_{Binding} $$

It would be interesting to study and compare quark-antiquark masses, particle-antiparticle masses, and matter-antimatter masses.

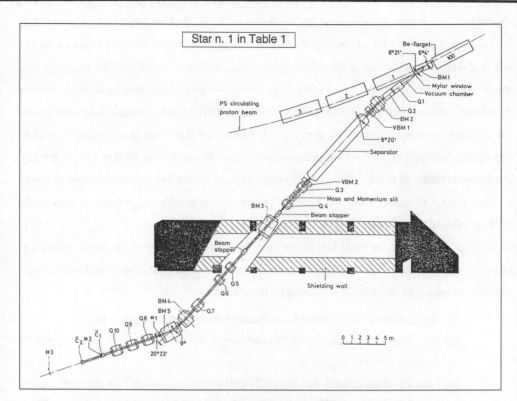

Star n. 1 in Table 1

The spectrometer to search for antideuterons (T. Massam et al., Il Nuovo Cimento, Vol. 39, p. 11 (1965)).

Melvin Schwartz, Tsung Dao Lee, Isidor I. Rabi in Erice (1968).
Tsung Dao Lee has pointed out that the CPT theorem at the Planck length loses its foundation.

Thirty years ago the key point was to establish if the antideuteron was there or not. In fact the collapse of the symmetry operators (C, P, CP, T), and the successes of the S-matrix in the description of strong interactions, did not allow anyone to take for granted that the nuclear binding forces had to be CPT invariant. In order to prove that the nuclear forces are CPT invariant it was necessary to reach the 10^{-8} level mentioned in § II.1-1 (Fig. II.2-1.2). Thirty years after the experimental proof of the existence of the first example of nuclear antimatter [49], this is understood in terms of an "effective" theory, the fundamental one (what we now think to be the fundamental one) being again in trouble.

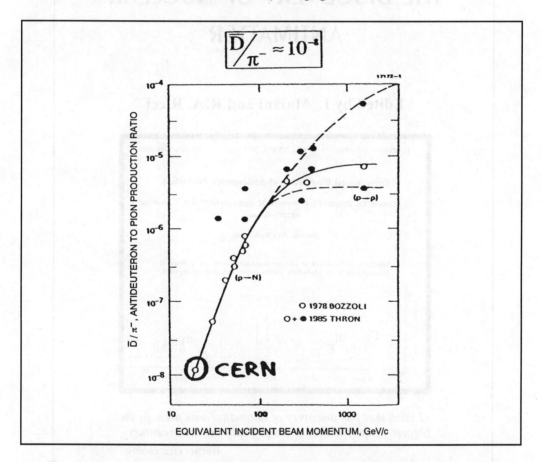

Fig. II.2-1.2: (Figure from Reference 110). The ratio $\overline{D}/\pi^-$ versus incident beam energy. Note the value 10^{-8} at production, in the energy range of the CERN discovery experiment.

In fact, in the framework of new physics at the Planck length, such as string theory, the CPT theorem has lost its foundations (T.D. Lee, 1995) [108]. On the other hand, in fields outside nuclear physics, there are experimental proofs that P, C, T, and CP Symmetries are violated but not the combination CPT. If experiments would indicate a violation of CPT, this would require a major reconstruction of our theories, since the present theoretical descriptions do not allow for CPT violation. From the rigorous theoretical standpoint, since

50

at the Planck scale the CPT theorem loses its foundation, the fact that the nuclear binding forces are CPT invariant rests only on firm experimental grounds, i.e. the existence of nuclear antimatter (Fig. II.2-1.3). If experiments would indicate a violation of CPT, this would require, as emphasized above, a major reconstruction of our theories, and therefore be of great importance.

Star n. 1 in Table 1

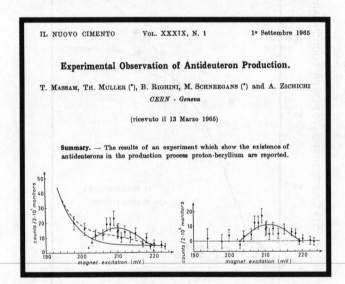

THE DISCOVERY OF NUCLEAR ANTIMATTER

Edited by L. Maiani and R. A. Ricci

IL NUOVO CIMENTO   VOL. XXXIX, N. 1   1º Settembre 1965

Experimental Observation of Antideuteron Production.

T. MASSAM, TH. MULLER (*), B. RIGHINI, M. SCHNEEGANS (*) and A. ZICHICHI

CERN - Geneva

(ricevuto il 13 Marzo 1965)

Summary. — The results of an experiment which show the existence of antideuterons in the production process proton-beryllium are reported.

«*I think that this discovery of antimatter was perhaps the biggest jump of all the big jumps in physics in our century*»

Werner Heisenberg

Fig. II.2-1.3: (Figure from Reference 111). The front page of the proceedings of the Symposium to celebrate the 30th anniversary of the Discovery of Nuclear Antimatter.

II.2-2   *The Third Lepton (1960-1975)*.   Stars n. 2^a, b, c in Table 1

During the sixties, the physics of hadrons appeared to be the only basic source of new knowledge. It was the "era" dominated by the "Bubble Chamber" technology.

During that "era", the first accurate measurement of QED radiative effects in a domain outside standard QED (electrons and photons) was performed at CERN [66-70].

In the following figures the 5‰ $(g-2)_\mu$ result (Fig. II.2-2.1) and the first high precision measurement (τ_μ) of the Fermi coupling [71] (Fig. II.2-2.2) are reported.

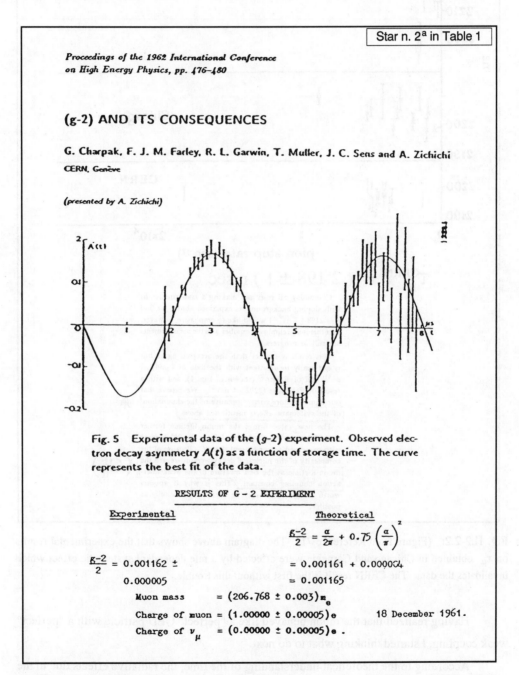

Proceedings of the 1962 International Conference on High Energy Physics, pp. 476–480

(g-2) AND ITS CONSEQUENCES

G. Charpak, F. J. M. Farley, R. L. Garwin, T. Muller, J. C. Sens and A. Zichichi

CERN, Genève

(presented by A. Zichichi)

Fig. 5 Experimental data of the $(g-2)$ experiment. Observed electron decay asymmetry $A(t)$ as a function of storage time. The curve represents the best fit of the data.

RESULTS OF G − 2 EXPERIMENT

Experimental

Theoretical

$$\frac{g-2}{2} = \frac{\alpha}{2\pi} + 0.75\left(\frac{\alpha}{\pi}\right)^2$$

$$\frac{g-2}{2} = 0.001162 \pm 0.000005$$

$$= 0.001161 + 0.000004$$
$$= 0.001165$$

Muon mass $= (206.768 \pm 0.003)\,m_e$

Charge of muon $= (1.00000 \pm 0.00005)\,e$

Charge of $\nu_\mu = (0.00000 \pm 0.00005)\,e$

18 December 1961.

Fig. II.2-2.1: (Figure from Reference 69). The first high precision measurement of QED radiative effects outside the (electron and photon) world.

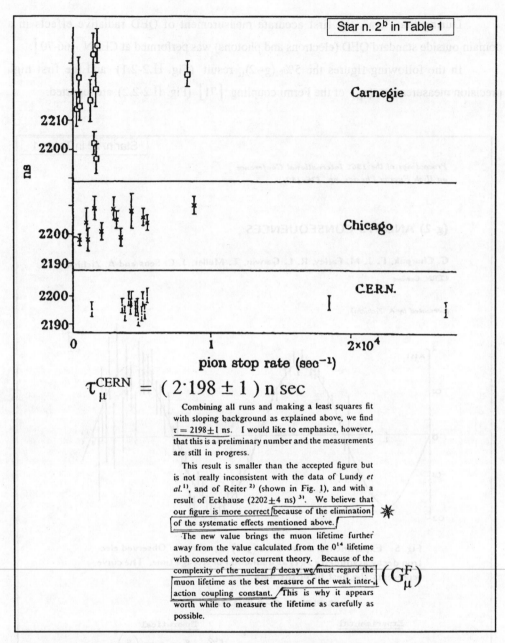

Combining all runs and making a least squares fit with sloping background as explained above, we find τ = 2198±1 ns. I would like to emphasize, however, that this is a preliminary number and the measurements are still in progress.

This result is smaller than the accepted figure but is not really inconsistent with the data of Lundy *et al.*[1], and of Reiter[2] (shown in Fig. 1), and with a result of Eckhause (2202±4 ns)[3]. We believe that our figure is more correct because of the elimination of the systematic effects mentioned above.

The new value brings the muon lifetime further away from the value calculated from the 0^{14} lifetime with conserved vector current theory. Because of the complexity of the nuclear β decay we must regard the muon lifetime as the best measure of the weak inter-, action coupling constant. This is why it appears worth while to measure the lifetime as carefully as possible.

Fig. II.2-2.2: (Figure from Reference 71). The diagram above shows that the experimental results on τ_μ obtained in Chicago and Carnegie were affected by a rate dependent systematic effect which invalidates the data. The CERN result is the first without this trouble.

Having realized that the muon behaved like a "perfect" QED particle with a "perfect" weak coupling, I started thinking what to do next.

According to the theoretical understanding of the time, the radiative effects due to the weak interactions of the muon had to diverge, but — as shown in Fig. II.2-2.1 — it was as expected by pure QED at the $\pm \frac{1}{2}$ % level of experimental accuracy. I realized that if the muon was heavier than the π–meson, i.e. if the π–meson had not the "ad hoc" mass, no-one would have ever seen a muon in our labs.

What about a third lepton? If its mass had been one GeV, would anyone have detected its existence? I started to discuss these matters with Weisskopf, Petermann and Cabibbo. The search for the third lepton took ten years of my life. This is documented in the volume [98] the front page of which is reported in Fig. II.2-2.3.

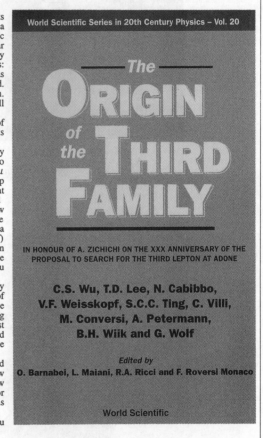

During the late fifties the great majority of physicists were fully engaged in strong interactions. Nevertheless, a high precision experiment on the anomalous magnetic moment of the muon, performed at CERN, gave a clear indication that the muon was a particle totally deprived of any sort of interaction but the electromagnetic and the weak ones: its magnetic anomaly was, within five parts in a thousand, as expected from pure QED with radiative corrections included. But the muon mass was 200 times heavier than the electron. Once the existence of the muon as a heavy electron was well established, what was the next step to be?

A large variety of proposals were presented, but none of the type thought of and thoroughly investigated in all its consequences at CERN by A. Zichichi.

In fact, (i) the idea for the existence of a new heavy lepton (HL) with its own leptonic number and coupled to its own neutrino, (ii) how to search for it (acoplanar $e\mu$ pairs), (iii) the construction of the first experimental set-up able to reject the high background levels, (iv) the proof that the best source for HL pairs was not $(\bar{p}p)$ but (e^+e^-), and (v) the experimental evidence that the search for this new heavy lepton using (e^+e^-) colliders could very well be achieved: all these were Zichichi's work for more than a decade, first at CERN and later, with the advent of the (e^+e^-) collider ADONE, at Frascati. The search for a new lepton coupled to its own neutrino was not an obvious and simple matter to deal with. It is easy to do the right thing once you have the right ideas.

That all these matters were not trivial ones is proved by the fact that all the papers published before 1970, the date of the first Frascati results, never considered the idea of the heavy lepton proposed and searched for by A. Zichichi during more than a decade of experiments; the key point was its best signature, a new effect: the production, above some threshold energy, of acoplanar $(e^{\pm}\mu^{\mp})$ pairs produced by time-like photons.

This volume has two main components: reports and testimonies. Both will allow the reader to know how this new field of physics was opened, how it gave rise to new technological developments (now still of great value for electron and muon detection), and how much work was needed for the "peculiar symmetry" to be so "short-lived".

Chien Shiung Wu

World Scientific Series in 20th Century Physics – Vol. 20

The

ORIGIN

of **THIRD**

the **FAMILY**

IN HONOUR OF A. ZICHICHI ON THE XXX ANNIVERSARY OF THE PROPOSAL TO SEARCH FOR THE THIRD LEPTON AT ADONE

C.S. Wu, T.D. Lee, N. Cabibbo, V.F. Weisskopf, S.C.C. Ting, C. Villi, M. Conversi, A. Petermann, B.H. Wiik and G. Wolf

Edited by
O. Barnabei, L. Maiani, R.A. Ricci and F. Roversi Monaco

World Scientific

Fig. II.2-2.3: Reference 98: the cover pages.

The best upper limit for the mass of the third lepton (Fig. II.2-2.4) was obtained using the Frascati (e^+e^-) collider searching for acoplanar $(e\mu)$ pairs. It was while searching exactly for this signature that the $HL^{\mp}$ was discovered at SLAC and called $\tau^{\mp}$. The first results on the $(e\mu)$ acoplanar search, at SLAC, were reported by G.J. Feldman at the EPS (1975) Conference in Palermo [112]. Fig. II.2-2.5 is a reproduction of page 1325 of my EPS closing lecture, where the interpretation of the SLAC acoplanar $(e\mu)$ pairs is cited in terms of the "old heavy lepton" searched for by the BCF group at Frascati [17].

Wolfgang Pauli and Chien Shiung Wu.

Chien Shiung Wu at Erice (1994).

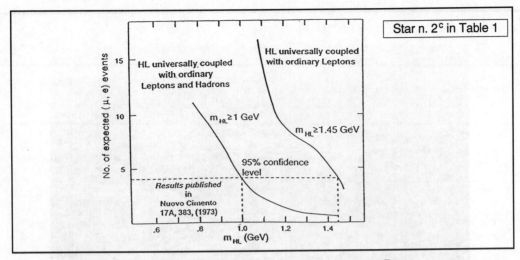

Fig. II.2-2.4: (Figure from Reference 17). The expected number of $(e^{\pm}\mu^{\mp})$ pairs vs. m_{HL}, i.e. the heavy lepton mass, for two types of universal weak couplings of the heavy lepton.

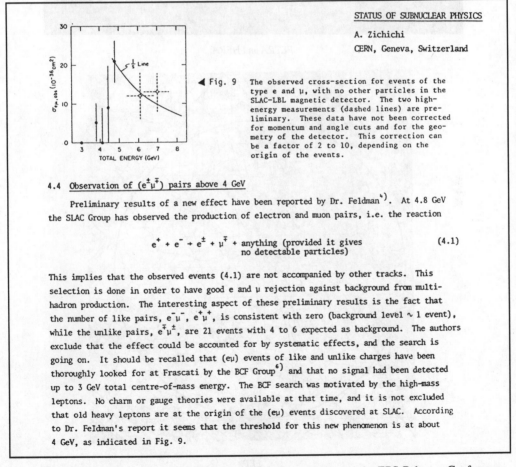

STATUS OF SUBNUCLEAR PHYSICS

A. Zichichi

CERN, Geneva, Switzerland

◀ Fig. 9 The observed cross-section for events of the type e and μ, with no other particles in the SLAC-LBL magnetic detector. The two high-energy measurements (dashed lines) are pre-liminary. These data have not been corrected for momentum and angle cuts and for the geometry of the detector. This correction can be a factor of 2 to 10, depending on the origin of the events.

4.4 Observation of $(e^{\pm}\mu^{\mp})$ pairs above 4 GeV

Preliminary results of a new effect have been reported by Dr. Feldman[4]. At 4.8 GeV the SLAC Group has observed the production of electron and muon pairs, i.e. the reaction

$$e^{+} + e^{-} \rightarrow e^{\pm} + \mu^{\mp} + \text{anything (provided it gives} \qquad (4.1)$$
$$\text{no detectable particles)}$$

This implies that the observed events (4.1) are not accompanied by other tracks. This selection is done in order to have good e and μ rejection against background from multi-hadron production. The interesting aspect of these preliminary results is the fact that the number of like pairs, $e^{-}\mu^{-}$, $e^{+}\mu^{+}$, is consistent with zero (background level ∿ 1 event), while the unlike pairs, $e^{\mp}\mu^{\pm}$, are 21 events with 4 to 6 expected as background. The authors exclude that the effect could be accounted for by systematic effects, and the search is going on. It should be recalled that (eμ) events of like and unlike charges have been thoroughly looked for at Frascati by the BCF Group[6] and that no signal had been detected up to 3 GeV total centre-of-mass energy. The BCF search was motivated by the high-mass leptons. No charm or gauge theories were available at that time, and it is not excluded that old heavy leptons are at the origin of the (eμ) events discovered at SLAC. According to Dr. Feldman's report it seems that the threshold for this new phenomenon is at about 4 GeV, as indicated in Fig. 9.

Fig. II.2-2.5: Reproduction of page 1325 of my closing lecture at the EPS Palermo Conference. The papers referred to are: 4 = Reference [112] and 6 = Reference [17] of the present review.

PETRA and HERA.

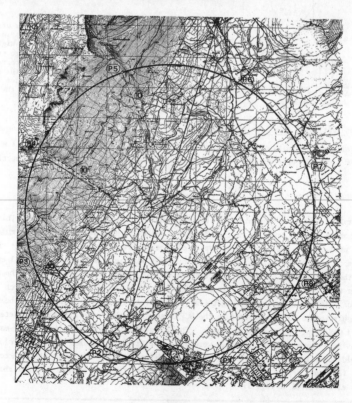

LEP.

The only missing element in the family structure of Fig. II.1.1 is the neutrino of the third family, called ν_{HL} in the first Frascati publication [17] and still to be discovered.

It is interesting to have a look at Fig. II.2-2.6 where the physics scenario which was at our disposal at energies above the ADONE (e^+e^-) collider limit is shown. The first step in this formidable scenario could have been discovered at Frascati if we had been allowed to run the ADONE collider (as we insistently requested) [113] above the nominal collider energy; a very small increase (just 3%) was needed for the J/ψ discovery, also known as the November revolution. If we had been allowed to run the ADONE collider above the nominal energy, the November revolution would have been originated by the INFN-Frascati Lab instead of BNL [114] and SLAC [115]. Also the 3rd lepton would have been discovered at the INFN-Frascati Lab, if the project to upgrade the ADONE collider [113] would not have been sabotaged by those who were campaigning against the Bologna-CERN group research programme to study narrow resonances and heavy leptons. Their motto was: Zichichi's group is searching for butterflies. The butterflies are shown in Fig. II.2-2.6.

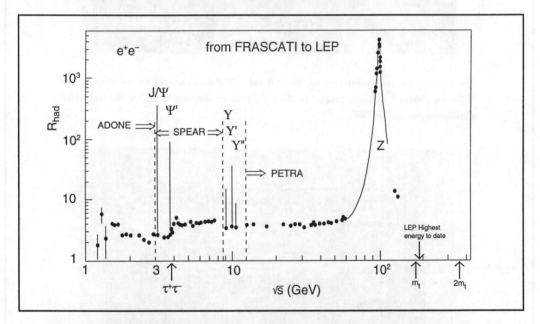

Fig. II.2-2.6: (Figure from Reference 116, updated with the latest LEP data). The "butterflies" as we know them now. The energy range from ADONE to LEP. On the vertical axis the ratio

$$\sigma_{had} = \frac{\sigma(e^+e^- \to \text{hadrons})}{\sigma(e^+e^- \to \mu^+\mu^-)} \quad ,$$

versus the centre-of-mass energy in the (e^+e^-) annihilation. Notice that the nominal ADONE energy was just 0.1 GeV below the J/ψ production threshold and 0.7 GeV below the threshold for the 3rd lepton. Notice also the energy gap between SPEAR and PETRA. In this gap there were the (b$\bar{b}$) states (Y, Y', Y") discovered at FERMILAB [117]. The maximum LEP energy is indicated as well as the energy threshold for the production of a single t and of a (t$\bar{t}$) pair.

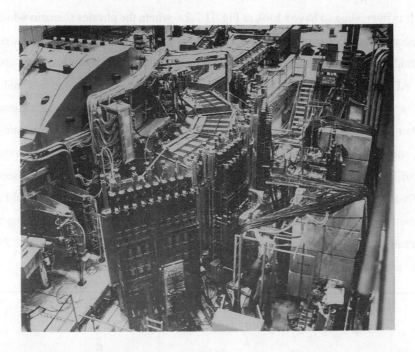

Two views of the experimental set-up, the ISR Split-Field-Magnet, used for the work on the "Effective Energy". Clearly visible, in the top figure, in the foreground are the large area scintillation counters for TOF.

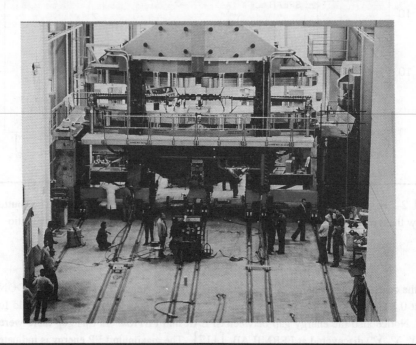

II.2-3 *The Effective Energy (1980-1984).* | Stars n. 3[a, b] in Table 1 |

As emphasised by G. Veneziano, the physics of strong interactions was characterized by two classes of phenomena, one of "static" nature, the other of "dynamic" nature. Both were affected by proliferation in the most fundamental component of this physics: its elementary particles.

The proliferation in the "static" sector of the strong interaction was the huge number of mesons and baryons [118]. As mentioned in § II.1-4[b], this multitude of states was reduced by an order of magnitude through the octets and decuplets of Gell-Mann and Ne'eman SU(3)$_f$ [99].

The proliferation in the "dynamic" sector was the multitude of final states produced by pairs of interacting particles, in strong, electromagnetic and weak processes:

Strong	EM	Weak
π p	γ p	ν p
K p	e p	
p p	μ p	
p n	e^+e^-	
$\bar{p}$ p		

It is the introduction of the effective energy which allowed one to put all the different final states on the same basis. This basis is the quantities measured in the multihadronic final states:

i) the average charged multiplicity; $< n_{ch} >$;

ii) the fractional energy distribution; $d\sigma / dx_i$;

iii) the transverse momentum distribution $d\sigma / dp_{t_i}$; etc. ..... .

The results are the universality features measured in all multihadronic final states, no matter what is the pair of interacting particles in the initial state.

The universality features are a QCD non-perturbative effect.

The first and basic step in this long "non-perturbative" QCD trip is the introduction of the effective energy. This new quantity came about by studying (pp) interactions at the

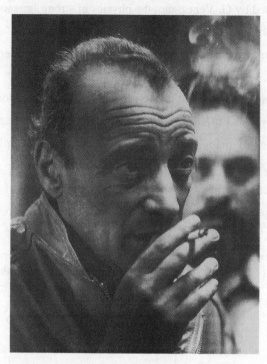

Vladimir N. Gribov

V. Gribov: I have three remarks.

Remark n. 1: When I read the paper "Evidence of the same multiparticle production mechanism in (pp) collisions as in (e⁺e⁻) annihilation", I realized that something very interesting had been found. In fact the introduction of the "Effective Energy" in the analysis of (pp) collisions at the CERN-ISR gave a totally unexpected result.

Remark n. 2: In the physics community there was a sort of gentlemen's agreement: please do not speak about results in contrast with the so much searched for gauge interaction to describe hadronic phenomena.

These "hidden" results were the hadronic systems produced in the interactions between pairs of hadrons; they were all different. Each pair of interacting particles, when producing systems consisting of many hadronic particles, had its own final state. No-one knew how to settle this flagrant contradiction. I wish I had the idea of the "Effective Energy".

Remark n. 3: Think of it. Even after so many years it appears to me a great achievement in physics.

A.Z.: Thank you, Volodya. I have a question. Why is QCD not able to "predict" the Universality Features?

V. Gribov: As you have emphasised in your lecture this is a non perturbative QCD problem. In order to answer your question I first need to understand confinement.

Vladimir N. Gribov.

CERN-ISR where it was proved that the set of final states produced at the ISR nominal energy of 62 GeV consisted of a sum of final states, each one having a different effective energy and the sum of all effective energies going from the lowest (few GeV) up to the highest ($\simeq$ 30 GeV) values allowed by our experimental set-up. These final states had properties like those produced in (e^+e^-) annihilation at the same "effective" ISR energies. The "nominal" ISR energy had to be corrected event by event. A quotation from the first paper where these results were reported is shown in Fig. II.2-3.1.

Star n. 3ª in Table 1

Introduction of the "**Effective Energy**"

EVIDENCE OF THE SAME MULTIPARTICLE PRODUCTION MECHANISM IN p-p COLLISIONS AS IN e^+e^- ANNIHILATION

M. Basile, G. Cara Romeo, L. Cifarelli, A. Contin, G. D'Alì, P. Di Cesare, B. Esposito, P. Giusti, T. Massam, F. Palmonari, G. Sartorelli, G. Valenti and A. Zichichi.

Physics Letters 92B, 367 (1980).

"The agreement between the momentum distributions obtained in e^+e^- annihilation and in p–p collisions suggests that the mechanism for transforming energy into particles in these two processes, so far considered very different, must be the same".

Fig. II.2-3.1: The first paper where the effective energy was introduced in the study of high energy (pp) interactions at ISR.

In order to check the validity of the effective energy down to the lowest value, the ISR collider was used at its lowest nominal energy: $(\sqrt{s})_{pp}$ = 30 Gev. This allowed a set of very low effective energies to be obtained using purely hadronic interactions. It was shown (Fig. II.2-3.2) that the fractional energies of the secondary particles produced in (pp) collisions had the same distribution as those produced in (e^+e^-) annihilation, at the same effective energy.

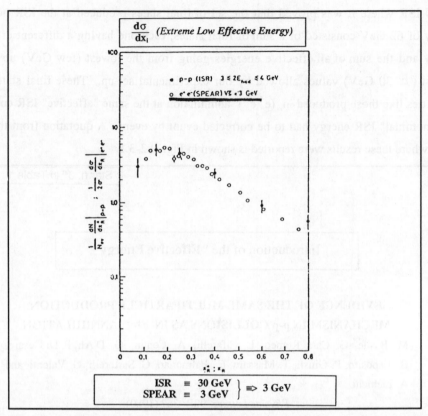

Fig. II.2-3.2: (Figure from Reference 119). The inclusive single-particle fractional momentum distributions $(1/N_{ev}) \cdot (dN/dx_R)$ in the interval $3 \text{ GeV} \leq 2E^{had} \leq 4\text{GeV}$ obtained from data at $\sqrt{s} = 30$ GeV. Also shown are data from MARK I at SPEAR.

Gerardus 't Hooft, the author and Gabriele Veneziano discussing the Effective Energy in QCD.

The fractional energy distributions of the multihadronic systems produced at the highest values of the "effective" ISR energies are compared with the same (e^+e^-) energies in Fig. II.2-3.3.

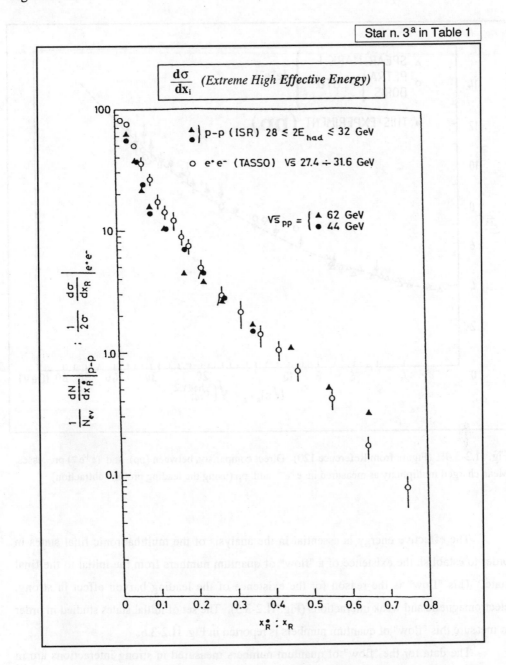

Fig. II.2-3.3: (Figure from Reference 120). These data compare (pp) and (e^+e^-) processes. The inclusive single-particle fractional momentum distributions $(1/N_{ev}) \cdot (dN/dx_R)$ in the interval $28 \text{ GeV} \leq 2E^{had} \leq 32 \text{ GeV}$ obtained from data at different $(\sqrt{s})_{pp}$ look like the data from TASSO at PETRA.

Another "universality" feature, the average charged multiplicity, $< n_{ch}>$, measured in (pp) and in (e^+e^-) interactions, is reported in Fig. II.2-3.4.

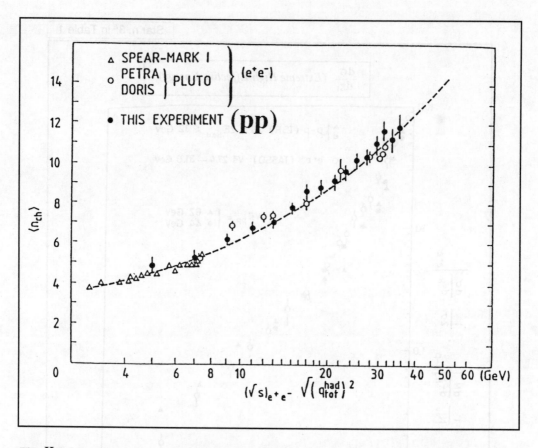

Fig. II.2-3.4: (Figure from Reference 120). Direct comparison between (pp) and (e^+e^-) processes. Mean charged multiplicity as measured in e^+e^- and pp (using the leading proton subtraction).

The effective energy is essential in the analysis of the multihadronic final states in order to establish the existence of a "flow" of quantum numbers from the initial to the final state. This "flow" is the reason for the existence of the leading baryon effect in strong, electromagnetic and weak interactions (Fig. II.2-3.5). The set of initial states studied in order to measure this "flow" of quantum numbers is reported in Fig. II.2-3.6.

The data for the "flow" of quantum numbers measured in strong interactions are in Fig. II.2-3.7, while those for the "flow" measured in electromagnetic and weak processes are in Fig. II.2-3.8.

Systematic study of the Leading Effect in
Strong EM Weak INTERACTIONS

**THE "LEADING"-BARYON EFFECT IN STRONG, WEAK,
AND ELECTROMAGNETIC INTERACTIONS**

M. Basile, G. Cara Romeo, L. Cifarelli, A. Contin, G. D'Alì, P. Di Cesare,
B. Esposito, P. Giusti, T. Massam, R. Nania, F. Palmonari, V. Rossi,
G. Sartorelli, M. Spinetti, G. Susinno, G. Valenti, L. Votano and A. Zichichi.

Lettere al Nuovo Cimento <u>32</u>, 321 (1981).

"This supports the idea that the "leading" phenomenon is generated by the quantum number «flow» from the initial to the final state".

Fig. II.2-3.5: Front page of the paper on the study of the leading effect with a sentence taken from the same paper [121].

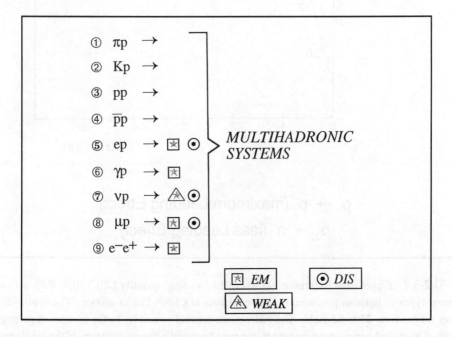

Fig. II.2-3.6: A list of initial states yielding multihadronic final states with the same properties, provided the "effective energy" is the same.

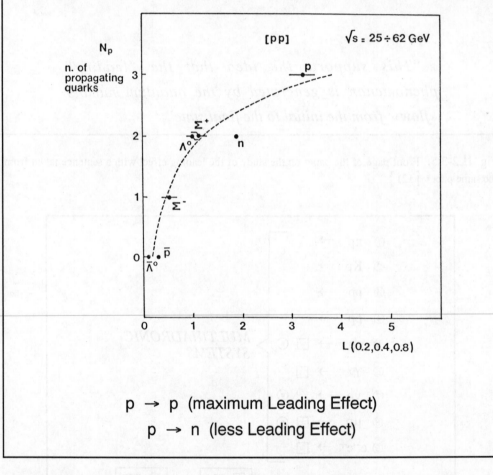

Fig. II.2-3.7: Figure from Reference 121 where the "leading" quantity L (0.2, 0.4, 0.8) derived for different types of baryons produced in (pp) collisions at CERN ISR is shown. The centre-of-mass energy ranges from 25 to 62 GeV. The hadrons are ordered according to the number of propagating quarks. The dotted curve superimposed is obtained by using a parametrization of the single-particle inclusive cross-section, $F(x) = (1 - x)^{\alpha}$, as described in section 3.

Flow of Quantum Numbers in <u>Electromagnetic</u> and <u>Weak Interactions</u>

326   M. BASILE, G. CARA ROMEO, L. CIFARELLI, A. CONTIN, G. D'ALÍ, ETC.

It should be noticed, as discussed in ref. (^{1}), that for Λ^0 production the energy dependence of L shows the same features as those observed for proton production. In fact, the three values of L obtained for the Λ^0 at ISR, at Fermilab, and at Cornell energies, suggest the following trend: the higher is the available energy, the lower is the value of L.

All the above results thus point out that it does/not matter whether the hadron interacts strongly, weakly, or electromagnetically: its «leading» effect is always present. ✳

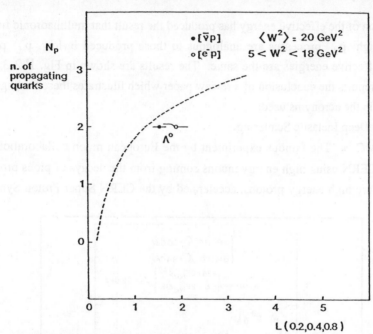

Fig. 2. – The «leading» quantity $L(0.2, 0.4, 0.8)$ of the Λ^0 produced in $(\bar{\nu}p)$ interactions at $\langle W^2 \rangle = 20$ (GeV)^2 and in (e^-p) interactions with $5 < W^2 < 18$ (GeV)^2. In this case the number of propagating quarks is two. The dashed curve of fig. 1 is also shown to guide the eye.

5. *Conclusions.* – In baryon-baryon interactions, the «leading»-baryon effect shows up very clearly in the x range $(0.2 \div 0.8)$. This «leading» effect is maximum when the final-state hadron is the same as the initial-state hadron. However, the «leading» effect is present even when the initial-state quantum numbers differ from those of the final state (for instance, when a proton becomes a Λ^0). As the difference between the initial- and the final-state quark composition increases, the «leading» effect decreases. This supports the idea that the «leading» phenomenon is generated by the quantum number «flow» from the initial to the final state. ✳

The «leading» baryon effect appears both in baryon-baryon and in lepton-baryon interactions. This means that a definite similarity must exist between processes in which a hadron is present in the initial state, no matter if the interaction is strong, weak or electromagnetic.

CERN
SERVICE D'INFORMATION
SCIENTIFIQUE

M. BASILE, et al.
14 Novembre 1981
Lettere al Nuovo Cimento
Serie 2, Vol. 32, pag. 321-326

Fig. II.2-3.8: Reproduction of a Figure from Reference 121 where the "quantum number flow" from the initial to the final state is observed in an electromagnetic process ($e^-p \to \Lambda^0 x$) and in a weak process ($\bar{\nu}p \to \Lambda^0 x$).

And now, another interesting point: the high transverse momentum myth (Fig. II.2-3.9).

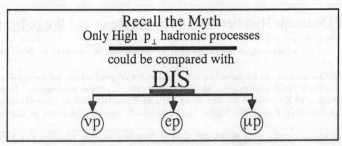

Fig. II.2-3.9: A synthesis of the high transverse momentum myth.

The introduction of the effective energy has produced the result that multihadronic final states produced in high $p_\perp$ processes are analogous to those produced in low $p_\perp$ processes, provided the effective energies are the same. The results are shown in Fig. II.2-3.10 while Fig. II.2-3.11 reports the conclusion of a review paper which illustrates these findings. A few words to explain the acronyms used:

DIS ≡ Deep Inelastic Scattering.

SPS/EMC ≡ The famous experiment by the European muon collaboration (EMC) performed at CERN using high energy muons coming from the decays of pions produced in collisions of very high energy protons, accelerated by the CERN Super Proton Synchrotron (SPS).

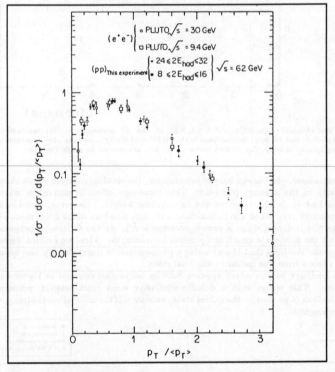

Fig. II.2-3.10: (Figure from Reference 122). Differential cross-section $(1/\sigma)\,[d\sigma/d\,(p_T/\langle p_T\rangle)]$ versus the "reduced" variable $p_T/\langle p_T\rangle$. These distributions allow a comparison of the multiparticle systems produced in e^+e^- annihilation and in pp interactions in terms of the "reduced" transverse momentum properties.

THE END OF A MYTH: HIGH-P_T PHYSICS

M. Basile, J. Berbiers, G. Cara Romeo, L. Cifarelli, A. Contin, G. D'Alì, C. Del Papa, P. Giusti, T. Massam, R. Nania, F. Palmonari, G. Sartorelli, M. Spinetti, G. Susinno, L. Votano and A. Zichichi.

Opening Lecture in Proceedings of the XXII Course of the
"Ettore Majorana" International School of Subnuclear Physics, Erice, Italy,
5-15 August 1984: "Quarks, Leptons, and their Constituents"
(Plenum Press, New York-London, 1988), 1.

"So far, the main picture of hadronic physics has been based on a distinction between high$-p_T$ and low$-p_T$ phenomena.

In the framework of parton model, high$-p_T$ processes were the only candidates to establish a link between

- *purely hadronic processes*
- *(e^+e^-) annihilations*
- *(DIS) processes.*

The advent of QCD has emphasized in a dramatic way the privileged role of high$-p_T$ physics due to the fact that, thanks to asymptotic freedom, QCD calculations via perturbative methods can be attempted at high$-p_T$ and results successfully compared with experimental data [1]. The conclusion was: we can forget about everything else and limit ourselves to high$-p_T$ physics.

Being theoretically off limits, low$-p_T$ phenomena, which represent the overwhelming majority of hadronic processes (more than 99% of physics is here), have been up to now neglected. By subtracting the leading proton effects in order to derive the effective energy available for particle production and by using the correct variables, the BCF collaboration has performed a systematic study of the final states produced in low$-p_T$ (pp) interactions at the ISR and has compared the results with those obtained in the processes listed below:

Process	Data Sources
(e^+e^-)	SLAC, DORIS, PETRA
(DIS)	SPS/EMC
(pp)	ISR (AFS)
$(\bar{p}p)$	SPS Collider (UA1)
(e^+e^-)	PETRA/TASSO (leading subtraction)

Transverse physics

The results of this study [2-18] show that, once a common basis for comparison is found by the use of the correct variables, remarkable analogies are observed in processes so far considered basically different like

- *low$-p_T$ (pp) interactions*
- *(e^+e^-) annihilations*
- *(DIS) processes*
- *high$-p_T$ (pp) and ($\bar{p}p$) interactions*

This is how universality features emerge, and this is the basis to proceed for a meaningful comparison, i.e.:

    <u>first</u> *identify the correct variables to establish a common basis,*

    <u>then</u> *proceed to a detailed comparison*."*

---

* *The root of this new approach to the study of hadronic interactions goes back a long time to a proposal by the CERN-Bologna group: "Study of deep inelastic high momentum transfer hadronic collisions" PMI/com-69/35, 8 July 1969."*

Fig. II.2-3.11: Reproduction of the conclusions of a review paper [123]. | Star n. 3^b in Table 1

ISR ≡ Intersecting Storage Rings. The first proton collider ever built. Its maximum energy was 31 GeV per proton beam.

ISR (AFS) ≡ The AFS (Axial Field Spectrometer) experiment using the colliding protons of the ISR.

(UA1) ≡ The experiment by Rubbia et al. to study high energy $(p\bar{p})$ collisions using the $Sp\bar{p}S$ (Super Proton antiProton Synchrotron) collider at CERN.

PETRA ≡ The (e^+e^-) collider built at DESY.

TASSO ≡ The experimental set-up built at DESY to study the final states produced in (e^+e^-) annihilation of energy (at that time) the highest in the world.

And now the question arises: is there a difference somewhere, in some corner, between the multihadronic final states produced in all these different processes?

The answer is "yes" and it refers to the multiplicity distributions; the jets in (pp) collisions have a strong component of gluonic origin, while the jets produced in (e^+e^-) annihilations are mostly originated by quarks. The data in Fig. II.2-3.12 indicate that the multihadronic states (jets) induced by gluons are flatter than those produced by quarks.

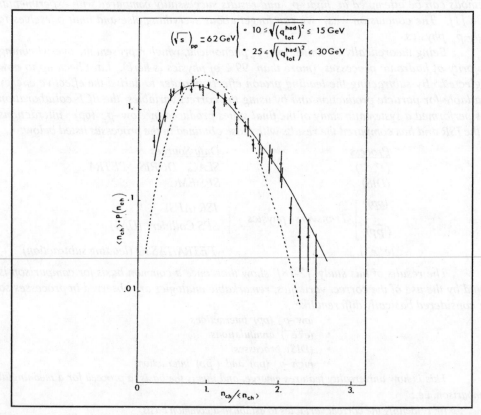

Fig. II.2-3.12: These distributions compare quark-induced jets (dotted line) with gluon-induced jets (full line). The $n_{ch}/\langle n_{ch}\rangle$ distributions measured in (pp) interactions, for two different intervals of $\sqrt{(q_{tot}^{had})^2}$ (open circles: 10 GeV $\leq \sqrt{(q_{tot}^{had})^2} \leq$ 15 GeV; solid circles: 25 GeV $\leq \sqrt{(q_{tot}^{had})^2} \leq$ 30 GeV). The solid curve is the best fit to the (pp) data [2.10]. The nominal (pp) c.m. energy was $(\sqrt{s})_{pp}$ = 62 GeV. The dotted line is the best fit to the (e^+e^-) data in the same energy range (Figure from Reference 119).

Question: is it possible to have a leading effect also in (e^+e^-) annihilation?

The answer came from PETRA, where the production of D^* from charm-quark in (e^+e^-) annihilation gave a clear leading effect:

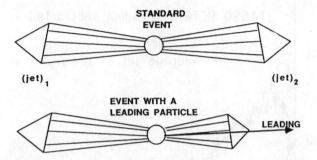

The data are shown in Fig. II.2-3.13 (not corrected for the leading effect) and in Fig. II.2-3.14 (corrected for the leading effect).

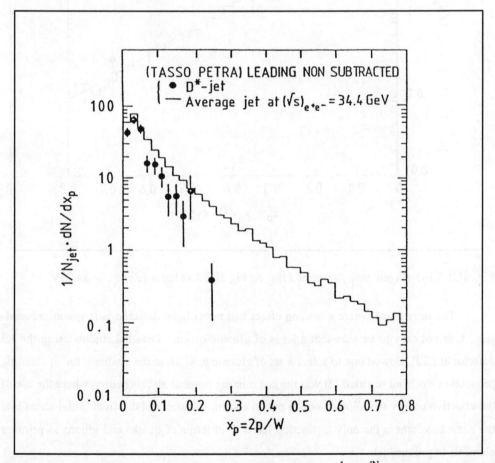

Fig. II.2-3.13: PETRA-TASSO collaboration. The distribution $\frac{1}{N_{jet}} \cdot \frac{dN}{dx_p}$ measured in standard jets and in jets containing a D^* at $(\sqrt{s})_{e^+e^-} = 34.4$ GeV (Figure from Reference 119).

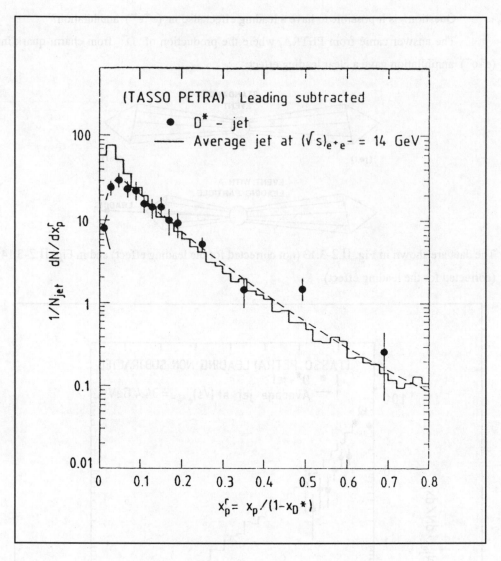

Fig. II.2-3.14: (Figure from Reference 119). As Fig. II.2-3.13 but at $(\sqrt{s})_{e^+e^-}$ = 14 GeV.

The only place where a leading effect had never been detected is in gluon-induced-jets. It is not easy to be sure that a jet is of gluonic origin. Detailed studies using the L3 detector at LEP allowed one to select a set of gluonic jets. Here the evidence for η' leading production has been reported. It was the last missing point in all this matter, where the use of the effective energy has allowed one to put an enormous variety of different initial states into the same box, where the only distinction left was in terms of quarks and gluons as primary elements to produce jets.

I will limit myself to report only one graph where η and η' production in gluon-induced-jets are compared (Fig. II.2-3.15). For a detailed report see Reference [125].

The interest of this finding is that the η'-meson, in order to be leading in a gluon-induced-jet, must have a strong coupling which can only be provided by its gluonic composition. It thus appears that the η' is the lowest pseudoscalar state generated by the fundamental force (QCD) which, as a by-product, produces the nuclear forces and therefore the π. In a sense, the π should have been the η'.

This brings me to a digression on the pseudoscalar mesons, fifty years later.

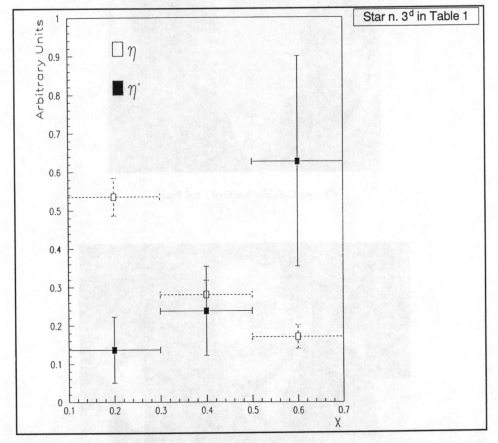

Fig. II.2-3.15: x–distributions for η and η' production, showing the leading effect (Figure from Reference 125).

II.2-4 *The pseudoscalar mesons and the Beppo particle (1947-1997).*   Star n. 3ᶜ in Table 1

Nuclear physics owes its origin to the Yukawa "meson" [58], experimentally discovered in 1947 by Lattes, Occhialini and Powell [2]. Fifty years later we know that the nuclear forces do not exist as fundamental forces. They are secondary effects of the fundamental force which is QCD.

What Yukawa was thinking is right, in terms of an "effective" theory, the fundamental one being drastically different. We now know that Yukawa's theory worked so well because the pion is much lighter than the nucleon.

The question thus arises: Why is the π-meson so light?

Giuseppe Occhialini and Cecil Frank Powell.

Giuseppe Occhialini.

The answer is threefold:

i) It could be thought that the π–meson should be light since it consists of a quark–antiquark $(q\bar{q})$ pair of the first family, which is made of very light quarks.

$$\pi \equiv q_I \bar{q}_I \quad \{ q_I \equiv \text{quark of the first family} \},$$

the q_I–mass being $\lesssim 10$ MeV. However there is a problem. In fact the confinement energy needed to keep $(q_I \bar{q}_I)$ together amounts to $\simeq 1000$ MeV, as proved by the mass of the nucleon

$$(q_I\, q_I\, q_I)$$

made of 3 quarks of the first family, all being nearly massless. So, the π–meson should be as heavy as the nucleon since the energy needed to keep quarks together is $\simeq 1000$ MeV.

ii) The quarks of the first family start as being nearly massless. They can therefore exist only as left or right states. This means that matter is chiral at the origin. What happens when we switch on QCD? This symmetry property (chirality) is not spoiled by the interaction between quarks and gluons. Why? Because the quanta of the non-Abelian force (QCD) are vectors. In fact, QCD is generated by a local invariance (the so-called gauge principle) (§ II.3-2).

iii) Chirality is spontaneously broken and since chirality–invariance is a global symmetry, its breaking must produce a physical effect, which is a massless particle, the Nambu-Goldstone-boson [126]. The π–meson is a (quasi perfect) Nambu-Goldstone-boson.

To sum up, the reason why the π–meson exists and is light, has to do with the existence of quarks which are matter fields, nearly massless, and therefore obeying chirality–invariance, a global symmetry property of nature. And it so happens that the strong force respects chirality–invariance because it is originated by a local invariance (for symmetry operations controlled by SU(3) in a fictitious space in three complex dimensions).

The π–meson is there to tell us that the original global symmetry of the matter fields (quarks) is spontaneously broken.

If it were not for the spontaneous breaking of chirality-invariance, the π–meson could not have 140 MeV mass and nuclear physics would not have started as the physics of a "fundamental" force of nature, having as typical range

$$R \simeq (140\,\text{MeV})^{-1} \simeq \text{one Fermi}.$$

The π–meson is not the quantum of the fundamental force (QCD). The quantum of this force is the gluon.

Yoichiro Nambu at Erice (1972).

IL NUOVO CIMENTO   Vol. XIX, N. 1   1° Gennaio 1961

A SYSTEMATICS OF HADRONS IN SUBNUCLEAR PHYSICS

YOICHIRO NAMBU

*The Enrico Fermi Institute for Nuclear Studies
and the Department of Physics, The University of Chicago, Chicago, Illinois*

(Received May 3, 1965)

1.

With the recognition that the SU(3) symmetry is the dominant feature of the strong interactions, the main concern of the elementary particle theory has naturally become directed at the understanding of the internal symmetry of particles at a deeper level. An immediate question that arises in this regard is whether there are fundamental objects (such as triplets or quartets) of which all the known baryons and mesons are composed. These fundamental objects would be to the baryons and mesons what the nucleons are to the nuclei, and the electrons and nuclei are to the atoms. If that was really the case, it would certainly precipitate a new revolution in our conceptual image of the world. At the moment we can only hope that the question will be answered within the next ten to twenty years when the 100 GeV to 1000 GeV range accelerators will have been realized.

Even now, the amusing and rather embarassing success of the SU(6) theory [1] lends support to the existence of those fundamental objects. It is embarassing because this is basically a non-relativistic and static theory, and we do not know exactly how this can cover the realm of high energy relativistic phenomena.

Putting aside those theoretical difficulties mainly associated with relativity, let us make the working hypothesis that there are fundamental objects which are heavy ($\gg 1$ GeV), though not necessarily stable, and that inside each baryon or meson they are combined with a large binding energy, yet moving with non-relativistic velocities. Though this might look like a contradiction, at least it does not violate the uncertainty principle in non-relativistic quantum mechanics since the range of the binding forces ($10^{-14} - 10^{-13}$ cm) is large compared

133

Field Theories with «Superconductor» Solutions.

J. GOLDSTONE

CERN - Geneva

(ricevuto l'8 Settembre 1960)

Summary. — The conditions for the existence of non-perturbative type « superconductor » solutions of field theories are examined. A non-covariant canonical transformation method is used to find such solutions for a theory of a fermion interacting with a pseudoscalar boson. A covariant renormalisable method using Feynman integrals is then given. A « superconductor » solution is found whenever in the normal perturbative-type solution the boson mass squared is negative and the coupling constants satisfy certain inequalities. The symmetry properties of such solutions are examined with the aid of a simple model of self-interacting boson fields. The solutions have lower symmetry than the Lagrangian, and contain mass zero bosons.

1. – Introduction.

This paper reports some work on the possible existence of field theories with solutions analogous to the Bardeen model of a superconductor. This possibility has been discussed by NAMBU [1] in a report which presents the general ideas of the theory which will not be repeated here. The present work merely considers models and has no direct physical applications but the nature of these theories seems worthwhile exploring.

The models considered here all have a boson field in them from the beginning. It would be more desirable to construct bosons out of fermions and this type of theory does contain that possibility [1]. The theories of this paper have the dubious advantage of being renormalisable, which at least allows one to find simple conditions in finite terms for the existence of « supercon-

[1] Y. NAMBU: Enrico Fermi Institute for Nuclear Studies, Chicago, Report 60-21.

Does a meson which is made with quanta of a fundamental force exist?

Is this meson a pseudoscalar state?

Is this meson the lightest state produced by the fundamental force?

The answer is three times "yes", and this meson is the η'–particle. Its mass is nearly one GeV, like the mass of another particle, the nucleon (made of three light quarks)

$$\eta' \, (g\bar{g}) \implies \text{mass} \simeq 1000 \, \text{MeV}$$
$$N \, (qqq) \implies \text{mass} \simeq 1000 \, \text{MeV} \; .$$

The reason being that a large fraction of the mass is due to confinement.

In fact,

$$\text{the mass of a gluon: } m(g) = \text{zero}$$
$$\text{the mass of a quark: } \quad q_I \lesssim 10 \, \text{MeV}$$

and

$$3q \implies \lesssim 30 \; \text{MeV} \implies 938 \, \text{MeV (proton)}.$$
$$2g \implies \lesssim \text{zero MeV} \implies 958 \, \text{MeV} \; (\eta').$$

Thus the mass of the lightest pseudoscalar particle made with two quanta of the fundamental force of nature (whose secondary effects produce nuclear physics) is as heavy as the "nucleon".

Fifty years after the particle imagined by Yukawa, we have now identified the lowest pseudoscalar state of what should be a particle made with quanta of the fundamental force acting between the constituents of a π–meson: gluons and quarks. This pseudoscalar state is the η' and this particle is as heavy as the heaviest known in 1947.

The η' typical range is therefore much smaller than that of the nuclear forces:

$$R \simeq [(1000) \, \text{MeV}]^{-1} \; .$$

For some time, after its discovery in 1964 [127], this pseudoscalar meson, the η', was called X^0, since its pseudoscalar nature was not established and there were mesonic states needed in the tensor multiplet of $SU(3)_f$. A meson with spin 2 cannot easily decay into 2γ and in fact the 2γ decay mode of the X^0 had not been observed, even when searched for down to a branching ratio level several times below that of the 2γ decay mode of the η^0, the well-known pseudoscalar neutral meson made of a quark–antiquark pair. This missing 2γ decay mode of the X^0–meson prevented the X^0–meson being considered as the singlet 9th member of the pseudoscalar $(q\bar{q})$ $SU(3)$–flavour multiplet structure of Gell-Mann and Ne'eman.

Gerardus 't Hooft in Erice (1975).

Murray Gell-Mann in Erice (1972).

The discovery of the 2γ decay mode of the X^0-meson [128] gave a strong support to its pseudoscalar nature. However its composition in terms of a quark-antiquark pair remained unclear. In fact, if a meson is made of a $(q\bar{q})$ pair, since quarks carry electric charges, the 2γ decay must be easily allowed. As mentioned above, the branching ratios of the 2γ decay mode of the two heavy pseudoscalar mesons were quite different and the absolute widths of the three pseudoscalar mesons, $\Gamma(\pi^0 \rightarrow \gamma\gamma)$, $\Gamma(\eta^0 \rightarrow \gamma\gamma)$ and $\Gamma(X^0 \rightarrow \gamma\gamma)$ did not follow the theoretical expectations.

Another difficulty was the X^0-mass. If the X^0-meson had to follow the Gell-Mann-Okubo (quadratic) mass formula, the mixing angle needed for these two pseudoscalar mesons was very small because the X^0-mass is nearly one GeV, compared with the $\simeq 0.5$ GeV η^0-mass. This mixing, when compared with the $(\omega$-$\phi)$ mixing, also measured [129] to be large (as expected), was the smallest known in all meson physics [130].

By now, the pseudoscalar nature of the X^0-meson is accepted and this meson is designated with the symbol η'. The notation now used is:

i) η^8, to indicate the 8th component of the $(q\bar{q})$ content of the pseudoscalar meson $SU(3)_f$ multiplet.

ii) η^0, to indicate the $SU(3)_f$ singlet component of the pseudoscalar $(q\bar{q})$ system.

These two components, η^8 and η^0, are not enough to describe the η' composition. In fact, we think we know the reason why the $(\eta$-$\eta')$ mixing angle is so anomalously small, namely the large gluonic content of the η'.

In QCD, the η and η' have played a decisive role. In the early days there was the so-called η-problem [105]. The theory appeared to demand a pseudoscalar η as an isosinglet made of non-strange quarks, and an η' as an $(\bar{s}s)$ state. Consequently the η-meson had to be close to the pion mass and the η' mass had to be near the K mass. The fact that experiments gave a quite different picture was attributed to the ABJ anomaly [84, 85] by Gell-Mann, Fritzsch and Leutwyler [105] and finally explained as an instanton effect by G. 't Hooft [91, 93]. Instantons induce a strong coupling between the η' and the two gluon state, and give this state a high mass, both of which may explain why the total width of the η' is so much bigger than that of the η. And consequently why the $\gamma\gamma$ branching ratio of the η' is so small [131].

Concerning experiments, for a number of years many attempts have been made to find out the gluonic content of the η', for example via a comparative study of the radiative decays of the (J/ψ) into η and η'. However all the methods adopted so far were based on indirect evidence. Only recently the first direct evidence for a strong gluonic composition of the η'-meson has been discovered [125]. If the η' has a strong gluon pair component, we

Harald Fritzsch at Erice (1998).

R. Jackiw at Erice (1973).

Gerardus 't Hooft at Erice (1998).

Gabriele Veneziano at Erice (1998).

should expect to see a typical QCD non-perturbative effect: the leading production in gluon-induced jets. In fact the leading effect had been observed in all hadronic processes where some conserved quantum numbers flow from the initial to the final state. If the gluon quantum numbers flow from an initial state made of two gluons into a final state made of η', this meson should be produced in a leading mode when the initial state is made of gluons. This is exactly the effect which has recently been reported in the production of the η'-mesons in gluon-induced jets.

Fifty years after the original idea of Yukawa that the quantum of the nuclear forces has to exist, we have found that this meson, called π, has given rise to a fantastic development in our thinking, the last step being the η'-meson. But the pseudoscalar nonet of mesons has been for many years a big problem for QCD. To solve it, many theorists had to think and work hard. Let me name them: Callan, Dashen, Gell-Mann, Gribov, Gross, Fritzsch, Jackiw, Leutwyler, Polyakov, Vainshtein, Veneziano, Witten, Zakharov and, most importantly, Gerardus 't Hooft who was able to finally explain the mass, the width and thus the $\gamma\gamma$ branching ratio of the η', introducing the instantons in QCD.

On the occasion of the 50th anniversary of the discovery of the π-meson, we would like to draw attention to the impressive series of conceptual developments linked with this discovery:

i)   The existence of a global symmetry property: chirality;

ii)   The spontaneous symmetry breaking of this global symmetry;

iii) The ABJ anomaly;

iv)   The existence of a non-Abelian fundamental force (QCD) acting between the constituents of the π-meson (quarks and gluons) and being generated by the gauge-principle which does not destroy chirality–invariance;

v)   The existence of another property of the non-Abelian force (QCD): the instantons;

vi)   The fact that chirality–invariance can be broken in a non-spontaneous way, thanks to the instantons.

Global chirality–invariance, spontaneous symmetry breaking, anomalies, gauge principle for non-Abelian forces, instantons: all originated from the π-meson and reached the final step with the η'-meson. It should be noticed that nearly all the credit for the π discovery went to Cecil Powell, a great leader and a very distinguished physicist. But the contribution of Beppo Occhialini deserves a recognition from the physics community. Thus, 50 years later, we propose the following. We started with the nuclear forces where the π-meson has played a central role; fifty years later we have the fundamental force QCD acting between the π-constituents: quarks and gluons. In QCD the $(\eta-\eta')$ problem has been a challenge for experimental and theoretical physicists. The role played by the X^0-meson is

Star n. 3ᶜ in Table 1

The "pre-shower" technology implemented in the CERN experimental set-up for the study of the rare decay modes of the pseudoscalar and vector mesons (1963). On the rails the "neutron missing mass spectrometer". This is the first example of what is now "standard" in experimental subnuclear physics: very large acceptance detectors.

crucial. First, very few believed it could be a pseudoscalar meson. Its mass and its width were too big and there was no sign of its 2γ decay mode. Once the X^0 was established to be a pseudoscalar meson, its gluonic affinity was needed and this was finally understood thanks to an important QCD development: the instantons. This theoretical picture has been experimentally proved to be correct with the discovery of the leading η' production in gluon–induced jets.

To sum up, the η' represents the conclusion of the π–meson challenge, and the basic steps are:

1 - The X^0–meson is discovered.

2 - The 2γ decay mode of the X^0–meson is discovered. The X^0–meson becomes the ninth member of the pseudoscalar multiplet and is called η'.

3 - The η'–meson is theoretically understood as being a mixture of $(q\bar{q})$ with a strong gluonic component, thanks to the QCD instantons.

4 - The strong gluon content in the η'–meson is experimentally proved to be present.

Both the experimental and theoretical front contributed to the physics of the η'–meson. We would like to propose to the physicists who have contributed to the four basic steps quoted above, that the η'–meson be called the Beppo Particle, to celebrate the outstanding contributions of Beppo Occhialini to physics, his humanity, modesty and devotion to science.

Star n. 5^a in Table 1

II.2-5 *Unification of the Gauge Couplings and the SUSY threshold (1979-1991-1993).*

We should never forget that subnuclear physics is based on experimental results. Why do we care so much about phenomena — like the gauge coupling unification — which occur at inaccessible energies?

The answer is simple: we want to know, as reliably as possible, which is the energy threshold for supersymmetry to be experimentally detectable. If this energy is in the TeV range we have to wait for the new generation of supercolliders. This conclusion was reached by rudimentary approaches [57], where threshold effects and the running of the gaugino masses were ignored.

If there are good reasons to believe that the supersymmetry threshold can be at the Fermi scale then we can exploit the facilities now available.

The first time I realized that supersymmetry was a new degree of freedom needed to have a better convergence of the gauge couplings was in 1979 [54] when, with André Petermann, we were trying to work out how to compute where the energy level of the

IL NUOVO CIMENTO   VOL. 105 A, N. 8   Agosto 1992

The Simultaneous Evolution of Masses and Couplings: Consequences on Supersymmetry Spectra and Thresholds.

F. Anselmo([1]), L. Cifarelli([1])([2])([3]), A. Peterman([1])([4])([5]) and A. Zichichi([1])

([1]) *CERN - Geneva, Switzerland*
([2]) *Physics Department, University of Pisa - Pisa, Italy*
([3]) *INFN - Bologna, Italy*
([4]) *World Laboratory - Lausanne, Switzerland*
([5]) *Centre de Physique Théorique, CNRS-Luminy - Marseille, France*

(ricevuto il 2 Aprile 1992; approvato l'8 Luglio 1992)

Summary. — We use the Renormalization Group Equations to work out, at the one-loop level, the simultaneous evolution of all masses and couplings and show explicitly the self-consistency of the whole scheme. A thorough examination is performed of the light Supersymmetry threshold in the Minimal Supersymmetric extension of the Standard Model (MSSM). All fundamental quantities α_{GUT}, M_{GUT}, $\alpha_3(m_Z)$, $\alpha_2(m_Z)$, $\alpha_1(m_Z)$ (consequently $\sin^2 \theta(m_Z)$) are given in terms of the detailed spectrum of all particle and sparticle thresholds. Examples of Supersymmetry spectra are given as function of $\alpha_3(m_Z)$ and of the other essential parameters. The results of this study, where the evolution of masses is extended to all possible masses, confirm our previous conclusions on the EGM effect for the Supersymmetry threshold lower bound. Examples of the predictive power of our method are given.

PACS 11.30.Pb – Supersymmetry.

1. – Introduction.

This work is in the line of our previous study[1] on the light threshold for Supersymmetry breaking[2], where a range for the primordial parameter $m_{1/2}$ was derived by comparing the experimental value $\sin^2 \theta(m_Z)_{\text{exp}}$ with $\sin^2 \theta(m_Z)_{\text{th}}$, *i.e.* the two-loop-corrected minimal SUSY-$SU(5)$ theoretical prediction, without threshold effects. The difference between these two quantities was then accounted for by the light threshold contribution, thus establishing, by means of the light spectrum, the range of energy where $m_{1/2}$ should lie.

In our previous work[1] we introduced the EGM effect, *i.e.* the evolution of the gaugino masses, in the light threshold study for SUSY breaking. Here we extend the light threshold effects to all unification parameters like M_{GUT} and α_{GUT} and obtain entirely analytic solutions for the one-loop evolution equations of the gauge supersymmetry threshold could be. We realized that too much work was needed before getting a decent estimate. When, in 1991 [57], we learned that the threshold was claimed to be in the TeV range [57], I had fifty fellows engaged in the search for SUSY signal using the L3 detector at LEP. If this claim was serious we were wasting our time. This is why, in 1991, I decided to go back with A. Petermann, and review our notes on a series of detailed studies of the basic problems connected with the gauge couplings unification [55], in order to try and understand the lowest possible values for the supersymmetry threshold [132]. These studies include the two loops RGEs [133], the EGM effect [56], and the study of the gap (see § II.2-6) between the string unification scale and the gauge coupling unification energy level. These studies have brought to the conclusion that the threshold for a supersymmetric signal to be detected, could be below the Fermi scale [134] and in fact accessible with the present facilities [135, 136] and in particular at LEP [137].

The final result of all these papers is a model based on "no-scale-supergravity" with one and only one free parameter. This single parameter allows one to predict a possible spectrum of all SUSY particles [138, 139]. This "prediction" is by far more founded than the claim by others [57] that the threshold had to be in the TeV range.

This "one-parameter-supergravity" model has been obtained using the top-mass (m_t = 175 GeV) and the constraints from radiative electro-weak symmetry breaking, plus the condition of B_0 = 0 at the GUT Scale. This allows one to solve for $\tan\beta$ in terms of $m_{1/2}$ thus reducing all free parameters to only one, plus the sign of μ determined to be negative.

In our papers [138, 139] the gravitino mass was not explicitly mentioned. However it is possible, in our no-scale-one-parameter-supergravity model, to get

$$m_{3/2} \ll m_{1/2} ;$$

more precisely:

$$10^{-5} \text{ eV} \lesssim m_{3/2} \lesssim 10^{-3} \text{ eV} .$$

In standard supergravity models, the generally accepted result is

$$m_{3/2} \simeq m_{1/2} .$$

In fact, the gravitino mass $m_{\tilde{G}}$ depends on the gauge kinetic function "f" and the usual expressions for "f" give the undesirable result

$$m_{1/2} \simeq m_{3/2} .$$

However, in the framework of no-scale-supergravity, the relation above becomes

$$m_{1/2} \simeq \left(\frac{m_{3/2}}{M_N}\right)^{1 - \frac{2}{3} q} M$$

where M_N is the Planck mass and M the unification scale mass. The value q such as to get the gravitino mass in the limits indicated above $(10^{-5}\,\text{eV} \lesssim m_{3/2} \lesssim 10^{-3}\,\text{eV})$ is perfectly allowed. It is interesting to recall that a light gravitino was obtained earlier by Barbieri, Ferrara and Nanopoulos in the framework of $N = 8$ supergravity with hierarchical supersymmetry breaking. The $N = 8$ difficulty has recently been overcome in the supersymmetric M-theory where it has been shown that $N = 8$ in $D = 4$ corresponds to $N = 1$ in $D = 11$. Thus our result with a very light gravitino in a one-parameter-supergravity model is perfectly consistent with the most recent picture of no-scale-supergravity and its breaking.

The restriction $m_{3/2} \ll m_{1/2}$ does not alter the spectra, but it changes the experimental signals to be searched for.

The most important ones are the hard photons from χ_1^0 decays.

In fact, in our model, the gravitino turns out to be the lightest supersymmetric particle (LSP) and the dominant decay of the lightest neutralino is

$$\chi_1^0 \rightarrow \gamma \tilde{G}$$

with $m_{\tilde{G}} < 250\,\text{eV}$. This opens a new final state to be searched for in (e^+e^-) interactions for SUSY signals: i.e. acoplanar photon pairs with nothing else but missing energy [140].

In Fig. II.1.3 the results of our analysis are shown in terms of the gauge couplings $(\alpha_1\,\alpha_2\,\alpha_3)$ unification. To the best of our knowledge the SUSY threshold could be as low as the Fermi Scale $(\simeq 10^2\,\text{GeV})$: this is where the gauge couplings change slope in the Figure.

II.2-6 *The Gap (1994)*. | Star n. 5^b in Table 1 |

Another important point is to establish to which extent the energy level of $\simeq 10^{16}$ GeV, where the three gauge couplings $\alpha_1\,\alpha_2\,\alpha_3$ converge towards a unique value, is really separated from the energy level of $\simeq 10^{18}$ GeV [141, 142], suggested by the relativistic quantum string theory. The reason why this is of interest to us is because the energy level could allow the choice of the correct supergrand unification gauge group. For example, $SU(5) \times U(1)$ would have no problem to fill the gap which exists between the two energy scales.

To identify the supergrand unified symmetry group has consequences on experiments at accessible energies, as for example the proton stability (τ_p), the existence of heavy neutrinos, the existence of monopoles and the supersymmetric particle spectrum.

A detailed analysis of the problems related to the existence of the gap is shown in Fig. II.2-6.1. The results of the analysis show that, by introducing the uncertainties in the experimental determination of the gauge couplings and on the basic masses which describe the GUT threshold, the gap can be filled.

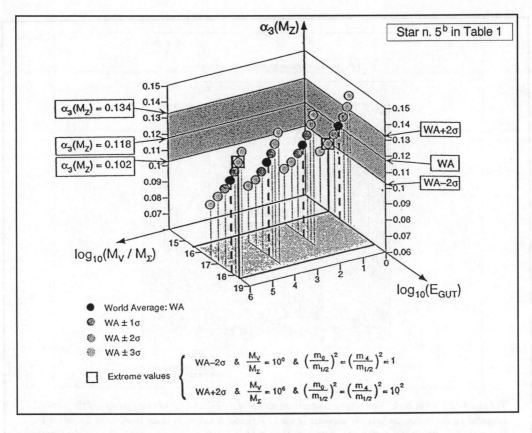

Fig. II.2-6.1: The dependence of E_{GUT} on $\alpha_3(M_Z)$ and on the ratio of the two crucial heavy-threshold masses, M_V / M_Σ. Note that the extreme value for E_{GUT} is above 10^{18} GeV (Figure from Reference 141).

II.3   *THE BASIC STEPS*

Introduction.

These past fifty years (1947-1997) have brought subnuclear physics to have a set of facilities — present and future — listed in the upper part of Fig. II.3.1. The ultimate step is the 200 TeV frontier: ELN.

What is now called LHC is in fact the 10% ELN (see § III).

In terms of new concepts and understanding, the past fifty years have brought us to the Standard Model with its more than 20 parameters and — fortunately — with other unsolved problems.

The basic steps (Fig. II.3.1) are five: i) the running of the gauge coupling and masses with energy; ii) the way a fundamental force is generated; iii) the physics of imaginary masses; iv) the flavour mixing and the breaking of the symmetry operators CP and T; v) the ABJ anomaly and instantons.

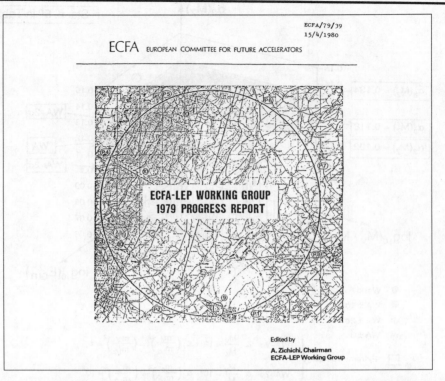

ECFA/79/39
15/4/1980

ECFA EUROPEAN COMMITTEE FOR FUTURE ACCELERATORS

**ECFA-LEP WORKING GROUP
1979 PROGRESS REPORT**

Edited by
**A. Zichichi, Chairman
ECFA-LEP Working Group**

The ECFA-LEP white book. This volume contains the results of the ECFA-LEP working group and the proposal, advanced by the author, to have the diameter of the LEP tunnel 5 metres in order to allow in the future the insertion of another proton collider in addition to the (e^+e^-) ring. This proposal was heavily criticised by the CERN Research Director of the time, but has been essential for the realization of LHC, as mentioned by L. Maiani (paper reproduced below).

PHYSICS WITH FUTURE COLLIDERS

EPS Conference, Bruxelles, 1995

LUCIANO MAIANI

*Department of Physics, University of Roma, P.le A. Moro 2, Roma, 00185
Istituto Nazionale di Fisica Nucleare, Italy*

The possibility to install the collider in the existing LEP tunnel has been crucial for the approval of the LHC project, which therefore owes much of its existence to the foresight of the physicists who asked for a tunnel with a much larger cross-section than needed at the time (see Ref.[1])[a]

1. ECFA-LEP Working group, 1979 Progress Report, ed. by A. Zichichi, ECFA/79/39, 1980.

[a] The point was very clearly stated by the ECFA-LEP Working Group[1] (Conclusions, p. 304): *It was noted that the choice of the cross-section of the LEP main tunnel takes into account the possibility of adding a proton machine later if needed. This was considered of great importance. It was even felt that a slight increase of the tunnel cross-section might be advisable and in any case provision should be made for accomodating the cryogenic equipment required for superconducting magnets.* The position was severely criticized at the time on the basis of the implied increase in cost (no surprise!) which, it was felt, was not tolerable and could endanger the LEP project itself

FACILITIES AND THE BASIC STEPS

FACILITIES

Present: GRAN SASSO - LEP - HERA - SLAC - TEVATRON - RHIC.
Future: (10%) ELN $\equiv$ LHC $\rightarrow$ TESLA $\rightarrow$ ELN.

BASIC STEPS

Present: the Standard Model; Future: the open problems.

① RGEs (α_i ($i \equiv 1, 2, 3$); m_j ($j \equiv q, l, G, H$)): $f(k^2)$.
- GUT ($\alpha_{GUT} \simeq \frac{1}{24})_m$ & GAP ($10^{16} - 10^{18}$) GeV.
- SUSY (to stabilize $\frac{m_F}{m_P} \simeq 10^{-17}$).
- RQST (to quantize Gravity).

② Gauge Principle (hidden dimensions).
— How a Fundamental Force is generated: SU(3); SU(2); U(1).

③ The Physics of Imaginary Masses: SSB.
— The Imaginary Mass in SU(2) × U(1) produces masses ($m_{W^\neq}$; m_{Z^0}; m_q; m_l), including $m_\gamma = 0$.
— The Imaginary Mass in SU(5) $\Rightarrow$ SU(3) × SU(2) × U(1) or in any – not containing U(1) – higher Symmetry Group $\Rightarrow$ SU(3) × SU(2) × U(1) produces Monopoles.
— The Imaginary Mass in SU(3)$_c$ generates Confinement.

④ Flavour Mixings & CP $\neq$, T $\neq$.
— No need for it but it is there.

⑤ Anomalies & Instantons.
— Basic Features of all Non-Abelian Forces.

Note:					
q	$\equiv$	quark and squark;	m_F	$\equiv$	Fermi mass scale;
l	$\equiv$	lepton and slepton;	m_P	$\equiv$	Planck mass scale;
G	$\equiv$	Gauge boson and Gaugino;	k	$\equiv$	quadrimomentum;
H	$\equiv$	Higgs and Shiggs;	C	$\equiv$	Charge Conjugation;
RGEs	$\equiv$	Renormalization Group Equations;	P	$\equiv$	Parity;
GUT	$\equiv$	Grand Unified Theory;	T	$\equiv$	Time Reversal;
SUSY	$\equiv$	Supersymmetry;	$\neq$	$\equiv$	Breakdown of Symmetry Operators.
RQST	$\equiv$	Relativistic Quantum String Theory;			
SSB	$\equiv$	Spontaneous Symmetry Breaking.			

Fig. II.3.1: The five basic steps in our understanding of nature. ① The renormalization group equations (RGEs) imply that the gauge couplings (α_i) and the masses (m_j) all run with k^2. It is this running which allows GUT, suggests SUSY and produces the need for a non point-like description (RQST) of physics processes, thus opening the way to quantize gravity. ② All forces originate in the same way: the gauge principle. ③ Imaginary masses play a central role in describing nature. ④ The mass-eigenstates are mixed when the Fermi forces come in. ⑤ The Abelian force QED has lost its role of being the guide for all fundamental forces. The non-Abelian gauge forces dominate and have features which are not present in QED.

Julian Schwinger at Erice.

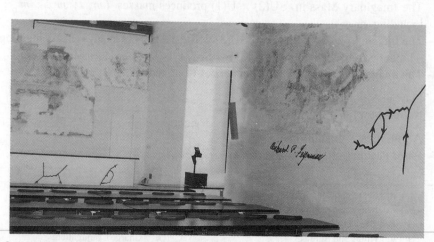

Last Supper "fresco" (dated around 1500) discovered while putting the iron reproduction of some original Feynman diagrams in the walls of the R.P. Feynman Lecture Hall in Erice. See Ref. [143].

An example of Feynman's contributions to the Discussions Sessions. Question by B. De Witt: «It was stated that gauge transformations depending on a single parameter generate local conservation laws. Would you distinguish local conservation laws from global conservation laws.»

R.P. Feynman: «If a cat were to disappear in Pasadena and at the same time appear in Erice, that would be an example of global conservation of cats. This is not the way cats are conserved. Cats or charge or baryons are conserved in a much more continuous way. If any of these quantities begin to disappear in a region, then they begin to appear in a neighbouring region. Consequently, we can identify a flow of charge out of a region with the disappearance of charge inside the region. This identification of the divergence of a flux with the time rate of change of a charge density is called a local conservation law. A local conservation law implies that the total charge is conserved globally, but the reverse does not hold. However, relativistically it is clear that non-local global conservation laws cannot exist, since to a moving observer the cat will appear in Erice before it disappears in Pasadena.»

Richard P. Feynman from Discussions at the International School of Subnuclear Physics; Erice, 27 August-7 September 1964, "Symmetries in Elementary Particle Physics", A. Zichichi (ed), Academic Press, New York and London, p. 380 (1964).

II.3-1 *The Renormalization Group Equations: RGEs*.

It is probably interesting to recall the basic conceptual developments. The first step was to realize that the radiative corrections to any electromagnetic process were logarithmically divergent. All divergencies could be grouped into two classes: one had the properties of a mass while the other had the properties of an electric charge. If these divergent integrals were substituted with the experimentally measured mass of the electron and with the experimentally measured value of its electric charge, all theoretical predictions could be made to be "finite". This procedure was called "mass" and "charge" renormalization. It was F.J. Dyson who proved, first the equivalence of the Tomonaga [144], Schwinger [145] and of the Feynman [146] "radiation" theory [147] and then showed that all divergencies arising from higher order radiative corrections could be removed by a consistent use of the idea of mass and charge renormalization [148]. All this was great, but the "mass" and "charge" renormalization appeared to be a "trick" [149].

An important step came about thanks to the work of E.C.G. Stueckelberg and A. Petermann [150] and of Gell-Mann and Low [151] who discovered that if the mass and the charge are made to be finite, they must run with energy. It was the starting point of what are now called the Renormalization Group Equations (RGEs) [152] which imply that all quantities, such as the gauge couplings (α_i) and the masses (m_j) run with k^2, the invariant quadrimomentum of a process. An interesting conceptual development is the invention, by M. Veltman and G. 't Hooft, of the "fractional dimension" [153, 154], which was found as a tool to renormalize non-Abelian gauge theories. With this tool the renormalization procedure became transparent. Finally, as reported by Sid Coleman in Erice during the 1973 School [155], the Feynman rules for gauge theories were settled [156] and the renormalization procedure in the presence of SSB really understood [153, 157, 158].

The literature on the renormalization group and related topics such as SSB is very large and I can only mention the memorable series of lectures by Sid Coleman in Erice [*Aspects of Symmetry - selected Erice Lectures*, Cambridge University Press, Cambridge, New York and Melbourne, 1985]. What remains out of all these formidable efforts is the running of the gauge couplings and of the masses. Starting from the high precision measurements at LEP, it turns out that all gauge couplings ($\alpha_1 \, \alpha_2 \, \alpha_3$) run towards a unique value $\alpha_{GUT} \simeq \frac{1}{24}$ at $E_{GUT} \simeq 10^{16}$ GeV. This number depends on what happens in the energy range from LEP to GUT. It is remarkable that the three gauge couplings converge towards the same value, despite the very many details not yet known. It is this "same value" which allows one to conclude that all gauge forces come from a unique grand unified origin (as shown in Fig. II.1.3). It should be emphasized that a nearly perfect unification can only be obtained if a new symmetry is introduced: SUSY, i.e. putting on the same basis

G. 't Hooft in "Under the Spell of the Gauge Principle"

«*Space and time are continuous. This is how it has to be in all our theories, because it is the only way known to implement the experimentally established fact that we have exact Lorentz invariance. It is also the reason why we must restrict ourselves to renormalizable quantum field theories for elementary particles. As a consequence we can consider unlimited scale transformations and study the behavior of our theories at all scales. This behavior is important and turns out to be highly nontrivial. The fundamental physical parameters such as masses and couplings (not constant) undergo an effective change if we study a theory at a different length and time scale, even the ones that had been introduced as being dimensionless. The reason for this is that the renormalization procedure that relates these constants to physically observable particle properties depends explicity on the mass and length scales used.*

It was proposed by A. Peterman and E.C.G. Stueckelberg[1], back in 1953, that the freedom to choose one's renormalization subtraction points can be seen as an invariance group for a renormalizable theory. They called this group the "renormalization group". In 1954 Murray Gell-Mann and Francis Low observed that the optimal choice of the subtraction depends on the energy and length scales at which one studies the system. Consequently it turned out that the most important subgroup of the renormalization group corresponds to the group of scale transformation. Later, Curtis G. Callan and independently Kurt Symanzik derived from this invariance partial differential equations for the amplitudes. The coefficients in these equations depend directly on the subtraction terms for the renormalized interaction parameters.

*The subtraction terms depend to some extent on the details of the subtraction scheme used. For gauge field theories Veltman and the author had introduced the so-called dimensional renormalization procedure. It turns out that if the subtraction terms obtained from this procedure are used for the renormalization group equations, these equations simplify. Furthermore there is a purely algebraic relation between the dimensional subtraction terms and the original parameters of the theory. This enables us to express the scaling properties of the most general renormalizable theory directly in terms of all interaction "**constants**" via an algebraic master equation. In deriving this equation one can make maximal use of gauge invariance. One can extend this master equation in order to derive counter terms for nonrenormalizable theories such as perturbative quantum gravity, but it would be incorrect to relate these terms to the scaling behavior of this theory, because here the canonical dimension of the interaction parameter, Newton's gravitational constant, does not vanish but is that of an inverse mass-squared.*»

1   *E.C.G. Stueckelberg and A. Peterman, Helv. Phys. Acta* <u>26</u> *(1953) 499.*

Advanced Series in Mathematical Physics
Vol. 19

UNDER THE SPELL
OF THE
GAUGE PRINCIPLE

G. 't Hooft

World Scientific

fermions and bosons [159].

This also allows one to stabilize the two scales where Spontaneous Symmetry Breaking (SSB) is needed. One of these scales is at $\simeq 10^2$ GeV (the Fermi scale), the other at $\simeq 10^{19}$ GeV (the Planck scale), and this is the origin of the hierarchy [160].

It happens that, to a first approximation, the gauge couplings $\alpha_1\ \alpha_2\ \alpha_3$ unify earlier and the problem of a gap between E_{GUT} and E_{Planck} arises, as discussed in § II.2-6. Using the relativistic quantum string theory, the unification energy level is shifted down, from the Planck scale to $\simeq 10^{18}$ GeV, thus reducing the gap to two orders of magnitude. If the gap was really there, a simple way out would be, for example, SU(5) × U(1) instead of SU(5) as the grand unification gauge force. In this case, the gauge couplings can be brought to converge at the same energy level as the one indicated by the string unification scale [21].

II.3-2   *The Gauge Principle and the Fundamental Forces.*

The second main theoretical step is that we finally understand how a fundamental force is generated [161]. It is through the requirement of a local invariance. Gravitational forces are generated by the requirement that we must be free to change a Lorentz frame at every space-time point. And this happens in real space-time, (Fig. II.1.2). But we need to imagine hidden dimensions as well. In these fictitious spaces with one, two and three complex dimensions (Fig. II.1.2) we can operate freely, provided that our operations obey the rules fixed by precise symmetry groups, such as U(1), SU(2) and SU(3). The freedom to operate in these fictitious spaces is the source of Quantum Electrodynamics (QED), Quantum Flavour Dynamics (QFD) and Quantum Chromodynamics (QCD). Thus, all fundamental forces appear to be generated by the same principle: the Gauge Principle [161], i.e. by requiring that the physics results have to remain the same when we operate in real space-time (gravity) or in fictitious spaces, each with different complex dimensions: three (QCD), two (QFD) and one (QED), following precise rules. As emphasized above, these rules are dictated by the appropriate symmetry groups which are: the Poincaré group for gravity and the SU(3), SU(2) and U(1) for the other forces, respectively. There is an important detail not to be forgotten and this brings us to the next point.

II.3-3   *The Physics of Imaginary Masses: SSB.*

QED and QFD are the result of a Spontaneous Symmetry Breakdown (SSB), occurring at the Fermi scale ($\simeq 10^2$ GeV) between the two gauge forces SU(2) and U(1). In other words neither QED nor the Fermi forces (QFD) are primary fundamental forces, as illustrated in Fig. II.3-3.1.

Peter Higgs in Erice with Sheldon Glashow (left) and Gerardus 't Hooft (1997).

This picture shows the three fellows who represent a synthesis of the main developments which brought us to the Standard Model: Sheldon Glashow was the first to identify the correct group for the Electroweak Interactions: SU(2) × U(1). Unfortunately, in this model, the masses of the vector bosons were not explained and created problems.   Peter Higgs was the first to have the correct idea on how to introduce masses without producing disasters. He discovered that via a spontaneous breakdown of local symmetry, massless Goldstone bosons become massive.   Gerardus 't Hooft was the only fellow who knew how to prove that the non-Abelian electroweak interactions described by the gauge forces SU(2) × U(1) could be shown to be renormalizable despite the spontaneous symmetry breaking (more precisely: the Higgs mechanism) needed in order to account for the masses of the weak bosons. Furthermore, the asymptotic freedom of QCD (the fact that the β–function has negative sign) and the consequent confinement at large distance scales, had in G. 't Hooft the fellow who first understood them. The experimental results opening these new avenues were: scaling in Deep Inelastic Scattering (DIS), discovered at SLAC, and the lack of free quarks established at CERN using the most powerful (pp) collider of the time.

Of course, Abdus Salam was the first to suggest that the Higgs mechanism should be used to solve the problem of the very massive weak boson, and Steve Weinberg was the first to propose that the Higgs mechanism had to be used for SU(2) × U(1). Concerning the Higgs mechanism, R. Brout and F. Englert were the first to show that via spontaneous breakdown of a local symmetry, not only Goldstone bosons, but massless vector bosons also become massive.

The conclusion of this long story is that a large number of theoretical breakthroughs were needed to arrive at the correct theoretical description of the fundamental forces of nature, as for example, the discovery by G. 't Hooft that local gauge invariance is not at all incompatible with finite masses for the vector particles at large distance scales. All one needs to do is to add physical degrees of freedom to the model that can play the role of the needed transverse field components. If these fields behave as scalar fields at short distances they do not give rise to any ghost problem there. At large distances they conspire with the vector field to describe the three independent components of a massive spin 1 particle.

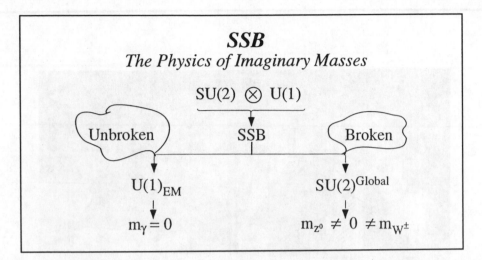

Fig. II.3-3.1: (Figure from Reference 5). A sketch of the Higgs mechanism, known as Spontaneous Symmetry Breaking (SSB). The gauge groups SU(2) × U(1), thanks to the introduction of a scalar with imaginary mass, are broken into two pieces. One survives the SSB and gives rise to an unbroken $U(1)_{EM}$ which is QED. The other part is broken and is the Fermi force.

These forces are the result of a mixing between two pure gauge forces SU(2) and U(1) which suffer a Spontaneous Symmetry Breakdown (SSB) at the Fermi scale, thus producing QED, QFD and all needed masses, i.e. those of the weak gauge bosons ($W^{\pm}$, Z^0) and those of quarks and leptons, including the zero-mass for the photon. The reason why the photon has zero mass is because $U(1)_{EM}$ survives unbroken despite SSB, as illustrated in Fig. II.3-3.1. The way SSB operates is illustrated in Fig. II.3-3.2.

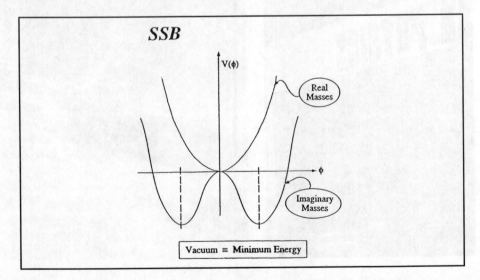

Fig. II.3-3.2: (Figure from Reference 5). The shape of the potential $V(\phi)$ is shown for two cases: real and imaginary mass. When the scalar field ϕ has imaginary mass, the vacuum is degenerate and two minima correspond to two values of ϕ which are different from zero. In other words $V(\phi)$ = minimum, but $\phi \neq 0$.

Steven Weinberg in Erice (1989).

Sheldon Glashow in Erice (1973).

Abdus Salam in Erice (1987).

All this is a basic pillar of the Standard Model, which would not be there if G. 't Hooft had not proved its renormalizability [156, 157]. In other words, the electro-weak process, i.e. the merging together of the electromagnetic and of the Fermi interactions, is a mathematical structure which produces finite theoretical predictions. Let us not forget that the starting points were two interactions with big problems: QED with all its divergences and troubles, emphasized by Landau with his poles [162], and the Fermi forces with their violation of unitarity at 300 GeV and with divergent results for any diagram except the zero order ones.

If masses are added "ad hoc" in the Lagrangian, as done by S. Glashow [163], R. Feynman [164] and M. Veltman [165], the theory is not renormalizable. On the other hand masses are needed: the weak bosons must be very massive. Moreover, $SU(2)_L$ alone cannot describe the weak forces since electrically charged and electrically neutral leptons are involved in the Fermi processes. In addition to $SU(2)_L$ another symmetry group is needed. The simplest addition to $SU(2)_L$ is $U(1)_{L, R}$ to take into due account the electric charge asymmetry which exists between charged (e^-), (μ^-) and neutral (ν_e), (ν_μ) leptons. The choice of the correct symmetry group $SU(2) \times U(1)$ is due to Sheldon Glashow [163].

What about the heavy vector bosons? S. Weinberg and A. Salam [166] had the right idea but they could not prove that the theory was renormalizable. The idea was to add in the Lagrangian a scalar with imaginary mass in the fundamental representation of $SU(2)$ and of $U(1)$. This allows one to have at one time very heavy weak bosons, $m_{W^\pm}$ and m_{Z^0}, (with $m_{W^\pm} \neq m_{Z^0}$), and an unbroken $U(1)_{EM}$ with a massless photon, m_γ = zero. It was G. 't Hooft who proved [157] that there is only one way to get a renormalizable theory with massive vector particles and this unique way is when the Lagrangian is exactly locally gauge invariant, in particular in its mass terms. This theoretical description of the electro-weak interactions is free of troubles [153, 167]: i) anomalies could be cancelled by requiring the same number of leptons and quarks in each family; ii) divergences could be cancelled by introducing the concept of "fractional dimensions" [153, 154]. All infinities exist if we insist on working in 4 dimensions. But if we use $(4 - \varepsilon)$ dimensions all integrals can be made finite. The important point is that, at the end, the physical quantities converge towards finite values when $\varepsilon \to 0$ and the number of dimensions goes back to 4 as in the real world.

Without this set of basic conceptual developments there would be no Standard Model.

II.3-4 *Flavour Mixing, CP violation and T breaking.*

Now a point which has no theoretical reason to exist.

As illustrated in Fig. II.3.4.1, quarks and leptons exist in four independent flavour

Nicola Cabibbo at Erice (1972).

Luciano Maiani at Erice (1975).

spaces, each having three dimensions. In the quark sector, one is for the up-type quarks (u, c, t); another one is for the down-type quarks (d, s, b). In the lepton sector, one is for the neutrinos and the other one for the negative leptons. Quantum Flavour Dynamics is not acting on each flavour separately. In the quark sector, the weak gauge coupling α_2 is given not only to a single quark flavour but also to a mixture. First, mixture was found to occur between the d quark and the s quark, and N. Cabibbo [168-171] used his famous angle θ_c to describe it. Then, when it was observed that there are no flavour changing neutral currents, a new mixing mechanism was required: Glashow, Iliopoulos and Maiani introduced the new quark c. With this quark present, the d-s rotation in flavour space led to the so-called GIM-mechanism [172]. Finally, with the discovery of CP violation (1964) by Christenson, Cronin, Fitch and Turlay [44], theoreticians concluded that two more quarks had to exist, t and b, and the Japanese physicists Kobayashi and Maskawa [173] succeeded in describing the mixing among the three quarks d, s and b using a matrix, the KM matrix. They found that one of their matrix elements could be complex, leading to CP violation. To be more precise there are in fact two mixtures: the up-type and the down-type, and the K.M. Matrix corresponds to the product of these two mixtures (A. Zichichi, Seminari di Fisica Superiore, Bologna University, 1977). Thus, in flavour space, the quark eigenstates, active for QFD Forces, are vectors not only along the axes indicated in Fig. II.3-4.1 (flavour non changing neutral currents) but also along rotated axes (flavour changing charged currents).

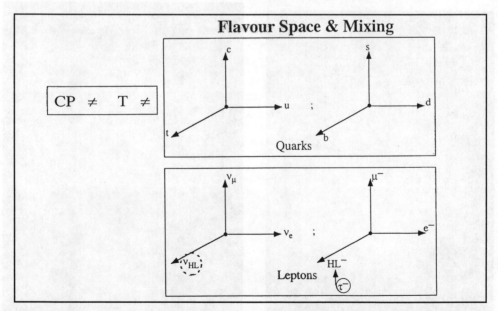

Fig. II.3-4.1: (Figure from Reference 5). The flavour space for quarks and leptons of the three families. As explained in the text, when the Fermi forces come in, the eigenstate of these forces are vectors not only along the axes (flavour non changing neutral currents) shown in the Figures. These vectors are also "rotated" (flavour changing charged currents) and therefore a mechanism mixing quarks of the "up" and "down" families becomes operative. Recently the same phenomenon was found to exist also in the lepton sector.

100

A view from inside of the Super-Kamiokande detector. Filling Water, January 1996. Data taking, April 1996 [174].

Masatoshi Koshiba at Erice (1998).

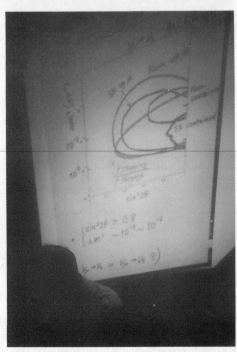

Masatoshi Koshiba at Erice (1998).

For many years the leptons appeared not to be affected by any mixing process. However, very recent results [174] obtained at SUPERKAMIOKANDE suggest that the flavour mixing mechanism is active also in the lepton sector.

No-one knows the origin of these mixtures nor the reason why these mixtures in the quark sector are so general as to cause a global symmetry breakdown of CP and T. All we can do is to measure all these quantities.

II.3-5 *Anomalies and Instantons.*

Finally, let me mention two great discoveries: anomalies and instantons.

- It is interesting to remark that Abelian QFT do not support instantons but

$$\lim_{k \to \Lambda_{Landau}} \alpha(k^2) \equiv \alpha_0 \equiv \infty \ .$$

This is the origin of all QED troubles such as the Landau poles [162].

II.3-5.1 *Anomalies.*

The anomalies correspond to quantum effects [175, 176].

The term "anomaly" is not so well-chosen since it refers to several different features in elementary particle theory. The term originated in QED where radiative effects were first discovered. It was introduced in order to describe quantum effects in Abelian QFT such as the "anomalous" magnetic moment of the muon (§ II.2-2).

- Non-Abelian QFT have chiral anomalies which must be cancelled, thus imposing severe conditions on the basic structures of the matter fields (example: the top quark needed in the third family).

- Anomalies exist also in Abelian theories, such as those needed to describe $\pi^0 \to \gamma\gamma$ [84, 85, 86]. They can thus be used to predict physical processes.

II.3-5.2 *Instantons.*

Let us now discuss the instantons which are basic features of non-Abelian QFT. The instanton [90, 91] is a solution of the classical field equations in Euclidean space-time. It is originated by the properties of the vacuum which is strongly coupled to the field quanta of a

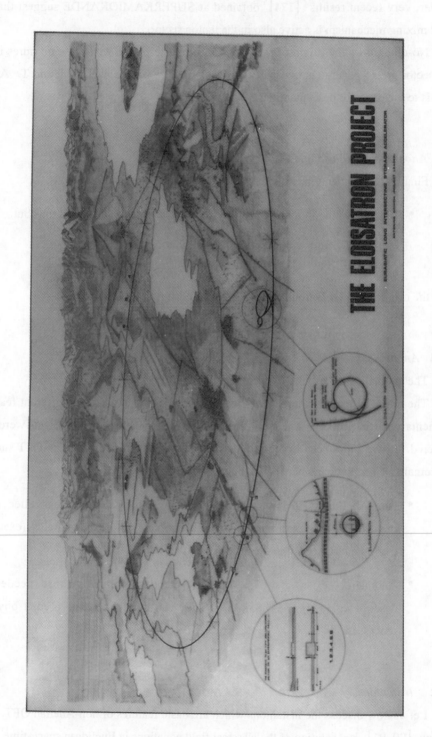

A view of the 300 Km ring positioned in the central part of Sicily.

given gauge force. In a quantized world the instanton corresponds to tunnelling effects in Minkowski space-time. These tunnelling effects are recognized in practice by the fact that they violate a global symmetry-law. There are two kinds of instantons, one for QCD and one for the QFD, the electro-weak forces.

In both cases, $SU(3)_c$ and $SU(2)_L$, i.e. QCD and QFD, the effects produced by the instantons can be understood in terms of the properties of the Dirac sea. In fact, the vacuum, made of fermions, has fermionic properties.

In QCD, these properties determine the "non-spontaneous", i.e. direct, breakdown of "chirality" invariance. This has allowed, as discussed in § II.2-4, to understand the behaviour of the η and the η' mesons $[177, 178, 179, 93]$.

In $SU(2)_L$ the effect of instantons is linked to the fact that the non-Abelian gauge force, QFD, acts only on left-handed states and instantons generate baryon number non-conservation, which is another $U(1)$ breaking.

Instantons typically have the effect of explicity breaking $U(1)$ symmetries.

III – THE ELN PROJECT.

III.1 *The Steps towards ELN.*

The Eloisatron project (ELN) started in 1979. The basic steps are shown in Fig. III.1.1. We have been able to build, in Italy, the superconducting cyclotron and 50% of the HERA superconducting dipole magnets[*] for the 800 GeV proton ring $[10]$. An intensive R&D project to study new subnuclear technologies has been implemented at CERN through LAA, which is part of the ELN strategy.

Since the very beginning, the conception of the Eloisatron project was based on a new way of tackling the problem of big projects. This new way is based on the following points:

1) Do not start to build a new laboratory if the work needed can be carried out within existing structures.

2) Involve the industry in the implementation of the project. And this also means that the R&D work should be done in the laboratories of the industries in close collaboration with the physicists.

These basic ELN steps have been the merging point for an effective collaboration between various universities, institutions and industries, as reported in Fig. III.1.2.

(*)   The original proposal was 100%, but German industries did not allow DESY to accept 100%.

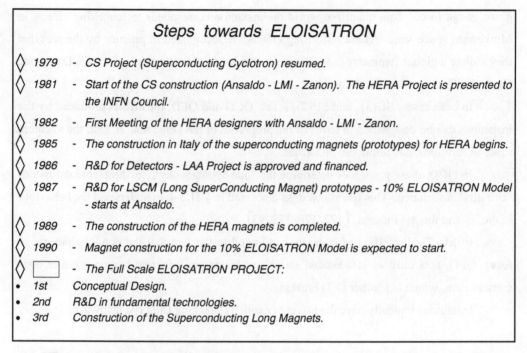

Fig. III.1.1: (Figure from Reference 180). The basic steps of the ELN strategy as reported when the HERA magnets were well installed at DESY.

Fig. III.1.2: (Figure from Reference 180). Universities, institutions and industries collaborating with ELN.

III.2   *The ELN Conceptual Design.*

The conceptual design of ELN has been performed by K. Johnsen and collaborators [180-187]. This study shows that there are no conceptual difficulties in the realisation of a (100+100) TeV collider able to work with a luminosity $L = (0.9 \div 1.8) \times 10^{33}$ cm^{-2}s^{-1}. The basic and lattice parameters of the collider are shown in Fig. III.2.1, and the collider layout in Fig. III.2.2.

BASIC PARAMETERS

Energy per beam	100 TeV
Number of bunches	39600 per beam
β-value at interaction point	1.25 - 0.6 m
Normalized emittance	$0.75\pi \times 10^{-6}$ m
r.m.s. beam radius at interaction point	1.25 - 0.9 x 10^{-6} m
Circulating current	16.43 mA
Particles per bunch	2.56×10^9
Beam-beam tune shift per crossing (with 6 active crossings)	1.67×10^{-3}
Bunch spacing	25×10^{-9} s
Stored beam energy	1.623×10^9 J
Luminosity (cm^{-2}s^{-1})	$0.9 - 1.8 \times 10^{33}$
Energy loss per turn due to synchrotron radiation	23.34 MeV
Radiated power (per beam)	385 KW
Power per unit length of one beam	1.89 W/m
Transverse em. damping time	1.2 h

LATTICE PARAMETERS

Length of period	200 m
Phase advance per period	$\pi/3$
Betatron wavelength	1200 m
Bending angle per normal period	4.7 mrad
Number of quads per period	2
Effective length of each quad	13.6 m
Nº of quadrupoles (without insertion)	2664
Maximum dipole field	10 Tesla
Bending radius	33356 m
Number of dipoles per normal period	12
Effective dipole length	13.1 m
Nº of dipoles	15984

Fig. III.2.1:   (Figure from Reference 180). Basic and lattice parameters of ELN.

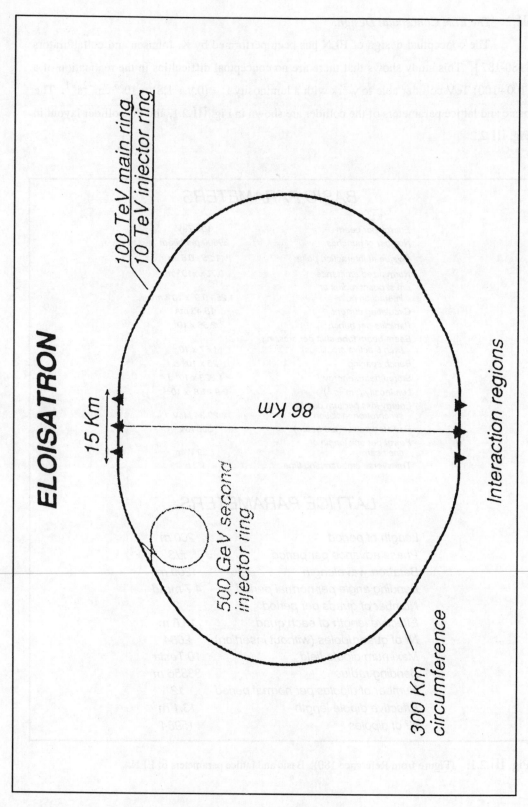

Fig. III.2.2: (Figure from Reference 180). The ELN collider layout.

A typical longitudinal cross-section of the ELN tunnel is shown in Fig. III.2.3 and a schematic longitudinal section of the ELN collider with the insertion of an experimental hall is shown in Fig. III.2.4. The schematic plan view of an experimental hall is shown in Fig. III.2.5. As shown in Fig. III.2.2 the experimental halls are (3 + 3) on two opposite sides. The 100 TeV ring is working as a 10 TeV injector to the same ring as shown in Fig. III.2.6. The ELN circumference is 300 Km.

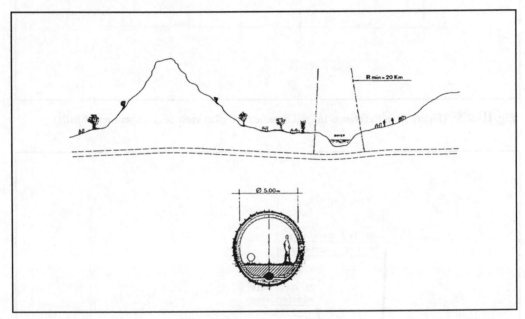

Fig. III.2.3: (Figure from Reference 188). A typical longitudinal cross-section of the Eloisatron tunnel.

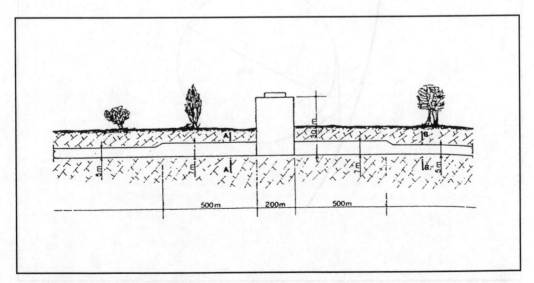

Fig. III.2.4: (Figure from Reference 188). The schematic longitudinal section of the ELN collider with an experimental hall inserted.

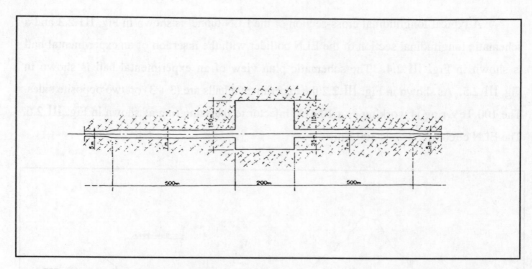

Fig. III.2.5: (Figure from Reference 188). The schematic plan view of an experimental hall.

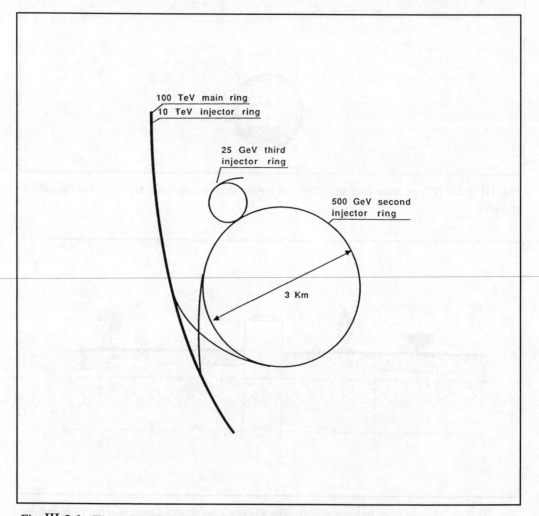

Fig. III.2.6: (Figure from Reference 188). The Eloisatron injection system.

III.3   *Results on Energy and Luminosity Limits.*

Our "machine" group has reached the following theoretical results:

- A collider designed for a luminosity L_1 cannot be operated at $100 \times L_1$ without spending a large fraction of the original cost of the collider main rings. Capability for high luminosity operation must be part of the initial design of a supercollider.

- To be specific:

  – The upgrade potential of SSC designed for $L = 10^{33}$ cm^{-2}s^{-1} will have been strongly limited by the already fixed design choices.

  – The central beam chamber of ELN should be designed for 10^{35} cm^{-2}s^{-1}.

- The 200 TeV (ELN collider) is practical with existing accelerator technology.

- A theoretical study of the maximum energy and luminosity levels attainable yields

$$E_{MAX} \quad = \quad (500 + 500) \text{ TeV}$$

and

$$L_{MAX} \quad = \quad 10^{36} \text{ cm}^{-2}\text{s}^{-1}$$

These ultimate ELN features (UELN) are shown in Fig. III.3.1 where other colliders are included for a comparative analysis.

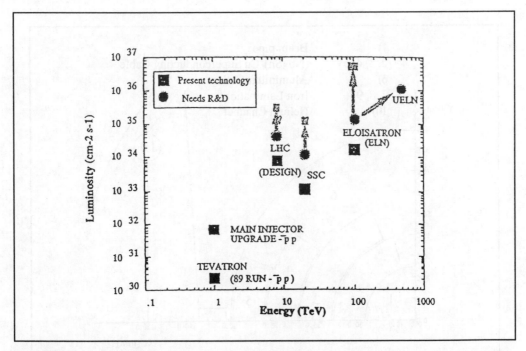

Fig. III.3.1: The ultimate energy and luminosity of ELN, compared with all other colliders in the world (Figure from Reference 180).

III.4 *Superconducting Dipole Magnet Studies.*

A series of dipole-magnet-design studies for the ELN collider has been made. The design goal is to reach a central field of

$$13.5 \text{ TESLA at } 4.2 \text{ K}.$$

Using Nb_3Sn superconducting material and cables, it is possible to reach 13.5 T with a quite good field uniformity:

$$(5.8 \div 13.5) \text{ T}, \quad \frac{\Delta B}{B^0} \leq 10^{-4}$$

$$(3 \div 5.8) \text{ T}, \quad \frac{\Delta B}{B^0} \leq (9 \div 1) \times 10^{-4}.$$

A dipole design is shown in Fig. III.4.1 for Scheme 3. Other schemes have been investigated [180-187] with 4-layer coils.

By choosing an average current density of

$$250 \text{ A/mm}^2 \text{ at } 14.28 \text{ T}$$

it is possible to get the needed Nb_3Sn cables with existing technology.

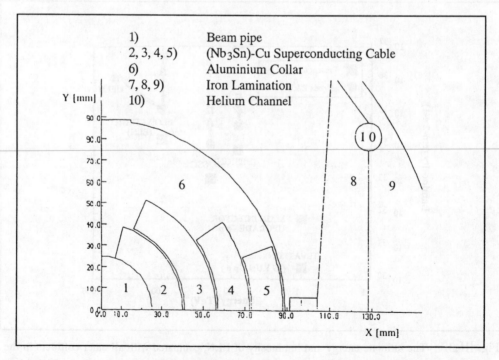

Fig. III.4.1: An example of detailed dipole design. One-quarter dipole cross-section. Scheme (3) for field uniformity calculation (Figure from Reference 180).

III.5   *R&D for Detectors able to work in the ELN conditions of Energy and Luminosities.*

The part of the ELN project which deals with R&D for detector technologies is called LAA [189-201].

Let me first mention the LAA basic points:

1. Radiation Hardness   •••
2. Hermeticity   •••
3. Rate Capability   •••
4. Momentum Resolution   •
5. Energy Resolution   •
6. Track & Space Resolution   •
7. Time Resolution   •
8. Particle Identification   •

The first three points are "new" since all previous experiments with existing colliders had never reached the levels of radiation hardness, hermeticity and rates typical of the ELN collider.

Note that hermeticity is an essential feature as the discovery potential of a detector lies essentially in its capability to see "missing" momenta. A (4π) detector has never been built. The challenge to study how to build a detector without holes in a (4π) solid angle coverage, able to handle high rates and to be radiation hard, has been our main (R&D) task during the past years of intensive work in the LAA project.

A synthesis of the main components needed for a (4π) detector working in the (20-200) TeV energy range is indicated in Fig. III.5.1, where the most important results obtained in each component in terms of numbers of discoveries, records, new developments and inventions are shown. It is impossible to report telegraphically and in a non-boring way the work of 200 specialists during ten years. The list of these results is reported in Tables 2 and 3.

Let me just say that it is thanks to this effort that many problems connected with the possibility of doing physics in the multi-TeV energy range, with high luminosity and rates, have been solved. If it were not for this component of the ELN project, i.e. LAA, we would not have been in a position to compete — here in Europe — with our American colleagues when planning the LHC detectors. In fact we are — in many areas of technological detector developments — ahead of them.

112

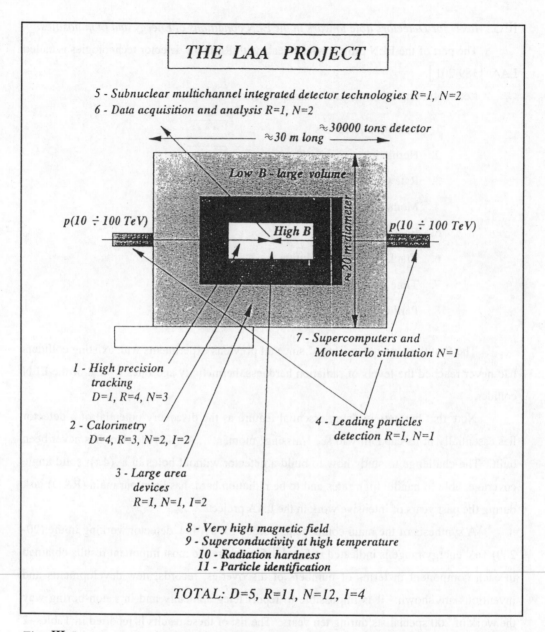

Fig. III.5.1: The basic scheme of a detector able to work at extreme values of Energy and Luminosity. D stands for Discovery; R for Record; I for Inventions; N for New developments. The numbers refer to the results obtained by the LAA project (Figure from Reference 180).

Table I.6.I. Main achievements of the LAA Project in terms of:
D=Discoveries, R=Records,
N=New developments, I=Inventions

	D	R	N	I
1. HIGH PRECISION TRACKING				
a) Gaseous detectors		2	1	
b) Scintillating fibres	1	2	1	
c) Microstrip GaAs			1	
2. CALORIMETRY				
a) High precision EM	2		1	
b) Compact EM+Hadronic	1	2		2
c) "Perfect" Calorimetry	1	1	1	
3. LARGE AREA DEVICES				
a) Construction				2
b) Alignment		1	1	
4. LEADING PARTICLE DETECTION		1	1	
5. SMIDT				
a) Microelectronics			2	
b) New, Radiation-resistant Technologies		1		
6. DATA ACQUISITION AND ANALYSIS				
a) Real Time Data Acquisition			1	
b) FASTBUS RISC computer		1		
c) Fine-grained Parallel Processor			1	
7. SUPERCOMPUTERS AND MONTECARLO SIMULATIONS			1	
TOTAL	5	11	12	4

Table 2: R&D for new detector technologies (LAA). The numbers of main results (D, R, N, I) are shown for each component (Table from Reference 201).

1. HIGH PRECISION TRACKING

a) Gaseous detectors
- Records:
  1. Rad-hard wire: ≫ 1 MRad
  2. Rate resistant: ≈ 2×10^6 particles/cm^2
- New developments:
  1. Mechanically reliable MWPC (MDM)

b) Scintillating fibres
- Discoveries:
  1. PMP
- Records:
  1. Smallest diameter sc. fibres: 15 μm
  2. Rad-hard fibres:≈ 1 MRad
- New developments:
  1. 1×1 mm^2 bundles with 900 fibres each

c) Microstrip GaAs
- New developments:
  1. GaAs particle detectors

2. CALORIMETRY

a) High precision EM
- Discoveries:
  1. EF
- New developments:
  1. New inorganic scintillators
  1. SSAC

b) Compact EM+Hadronic
- Discoveries:
  1. Intrinsic resolution better in Pb than U
- Records:
  1. Hadronic energy resolution: 27% $\sqrt{E}$
  2. Rad-hard fibres: >1 MRad
- Inventions:
  1. Spaghetti calorimeter
  2. e/π rejection by timing

c) "Perfect" Calorimetry
- Discoveries:
  1. CsI+TMAE adsorbed layer quantum efficiency
- Records:
  1. Highest efficiency photocathode (CsI+TMAE)
- New developments:
  1. Systematic studies of liquid Xe

3. LARGE AREA DEVICES

a) Construction
- Inventions:
  1. Blade chambers
  2. Gaseous pixel chambers

b) Alignment
- Records:
  1. Best length/angles/linearity measurements
- New developments:
  1. New instruments for length/angles/linearity measurement

4. LEADING PARTICLE DETECTION

- Records:
  1. Best way to cut silicon: < 50 μm cracks
- New Developments:
  1. Leading particle detector problems solved

5. SMIDT

a) Microelectronics
- New developments:
  1. All components of HARP done
  2. Read-out electronics for pixel detector

b) New, Radiation-resistant Tech.
- Records:
  1. Rad-hard amplifier: > 1MRad

6. DATA ACQUISITION AND ANALYSIS

a) Real Time Data Acquisition
- New developments:
  1. Test of commercial architectures with algorithms for feature extraction

b) FASTBUS RISC computer
- Records:
  1. Highest computer power in a FASTBUS board: 50 VAX-equivalent

c) Fine-grained Parallel Processor
- New developments:
  1. ASP construction and test of chips

7. SUPERCOMPUTERS AND MONTECARLO SIMULATIONS

- New developments:
  1. Full MonteCarlo Chain

Table 3: The main achievements of the LAA project – Details (Table from Reference 201).

III.6 *Physics Scenarios and Montecarlo Simulation Studies.*

The bulk of soft processes (i.e. minimum bias events) in hadron-hadron collisions is interesting from different viewpoints:

- "leading" hadron and effective energy for multiparticle production;
- dynamics of strong interactions in the non-perturbative domain;
- overwhelming "background" in the search for new physics phenomena.

The present theoretical situation is paradoxical.

Rare processes (high-p_T jets, W/Z, etc.) are by far easier to describe than low momentum transfer processes which do not have so far a satisfactory theoretical treatment.

The description of soft processes is considerably complex since two colliding hadrons are:

- composite;
- extended.

PYTHIA (perturbative) and QGSM (non-perturbative) models have been compared in pp and p$\bar{\text{p}}$ inelastic collisions [202-206] at:

$$\sqrt{s} \;=\; 62\,\text{GeV} \qquad [\text{ISR}]$$
$$\sqrt{s} \;=\; 630\,\text{GeV} \qquad [\text{Sp}\bar{\text{p}}\text{S}]$$
$$\sqrt{s} \;=\; 1.8\,\text{TeV} \qquad [\text{Tevatron}]$$
$$\sqrt{s} \;=\; 16\,\text{TeV} \qquad [\text{LHC}]$$
$$\sqrt{s} \;=\; 40\,\text{TeV} \qquad [20\%\ \text{ELN}]$$
$$\sqrt{s} \;=\; 200\,\text{TeV} \qquad [\text{ELN}]$$

The comparison has been done in terms of inclusive distributions of multiplicity and other variables, such as x_F, rapidity, etc., for different secondary hadrons, in particular for light and heavy baryons [203, 204].

We have focused our simulations on heavy Higgs search at 16, 40 and 200 TeV energies [202, 205, 206].

$$\text{Heavy Higgs search at}$$
$$\sqrt{s} \;=\; 16, 40, 200\,\text{TeV}$$
$$(M_H > 2\,M_Z).$$

The main production mechanisms investigated are:

$$gg \;\rightarrow\; H^0$$
$$W^+W^- \;\rightarrow\; H^0.$$

The results are reported in Fig. III.6.1. How a heavy Higgs can be observed in a real (LAA-like) set-up has been extensively studied. Let me show the results, in terms of signal-

116

to-background, for the case of heavy Higgs production, $M_H = 1$ TeV ($m_t = 170$ GeV), at three total c.m. energies: $\sqrt{s} = (16, 40)$ TeV (Fig. III.6.2) and $\sqrt{s} = 200$ TeV (Fig. III.6.3).

To sum up, extensive studies have been performed, by extrapolating all we know (QED, QFD, QCD) to extreme energies, in order to have a quantitative estimate of all possible "background" sources. Leading effects in high and heavy flavours production have been computed.

There are good reasons to believe that the lightest Higgs particle should be below 10^2 GeV. However other Higgs are needed heavier than the lightest one.

If a Higgs with $M_H \simeq 1$ TeV is there, there is no question that ELN is the best collider as shown in Fig. III.6.3.

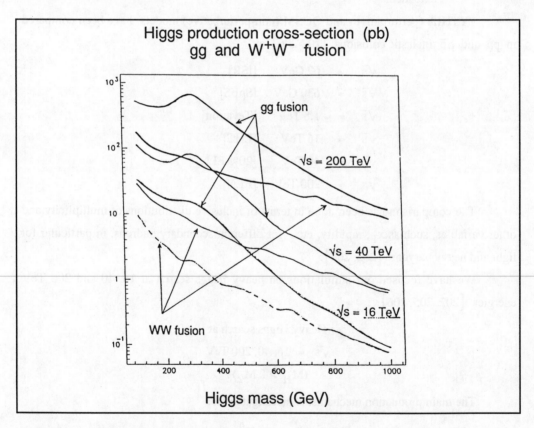

Fig. III.6.1: Higgs production cross-section (pb) (versus Higgs mass) at $\sqrt{s} = (16, 40, 200)$ TeV following two different production processes: (gg) and (WW) fusion (Figure from Reference 180).

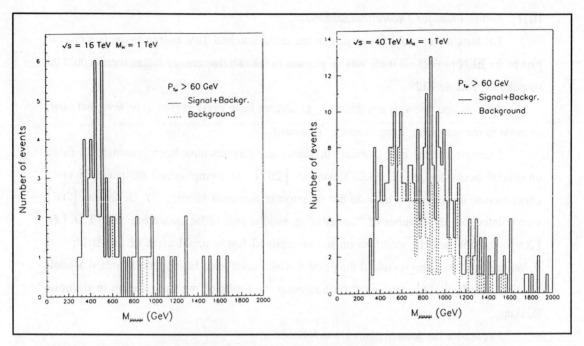

Fig. III.6.2: Heavy Higgs: signal/background at $\sqrt{s} = (16, 40)$ TeV (Figure from Reference 180).

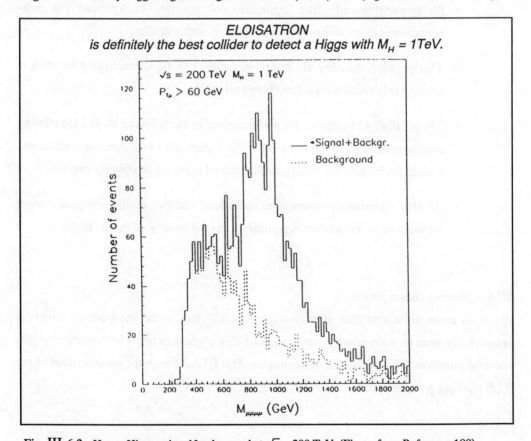

Fig. III.6.3: Heavy Higgs: signal/background at $\sqrt{s} = 200$ TeV (Figure from Reference 180).

III.7   *Future Plans for Theoretical Studies.*

To think about problems arising in the multi-hundred TeV energy range is a crucial part of the ELN project. If there was no physics in the extreme energy range, there would be no point to promote ELN.

This, as you know, is not the case. In fact we have every reason to believe that now, as never in the past, a big jump in energy is needed.

Examples of the new frontiers in subnuclear physics have been presented in Erice on several occasions by: i) S.L. Glashow [207] who emphasized the need for new experimental discoveries which do not fit into the Standard Model; ii) D. Gross [208] who pointed out the number of "fantastic" questions still to be answered; iii) T.D. Lee [209] who focused the attention on the unexplored fundamental areas of our field; iv) F. Wilczek [210], who reviewed the physics which must exist beyond the Standard Model; v) G. 't Hooft [211], who went on analysing the limits of our imagination in rigorous thinking.

In practice the present plans are as follows:

- Phenomenologically, the signature and possible background for non-perturbative electro-weak processes deserve further studies.

- The question whether B- and L-violation can be disentangled in such a multiparticle process is far from being solved.

- The possibility of large σ for non-perturbative electro-weak B- and L-violating processes needs further work. The final goal is to compute the total cross-section for B-violation [212] to exponential accuracy at arbitrary energies.

- Another interesting problem is to understand whether similar non-perturbative interactions also contribute significantly to total cross-sections in QCD.

III.8   *Working Group Structure.*

In order to be sure that subnuclear physics will stay in the forefront of scientific research, we need to study now the multi-hundred TeV physics in terms of theoretical goals, machine problems and new detector technologies. Fig. III.8.1 shows the present status of the ELN working group structure.

WORKING GROUPS

1. **MACHINE**
   - Theoretical: optics, beam dynamics etc.
   - R&D for the highest levels of luminosity and energy.
2. **DETECTORS**
   - R&D on Super Detectors capable of working at the highest energies and luminosities.
3. **PHYSICS**
   - Montecarlo Simulations at the highest Energy ($\gtrsim$ 200 TeV) and Supercomputers.
   - Theoretical studies.
   - World Data Base and Networking.

Fig. III.8.1

III.9 *Why 200 TeV now.*

There are two possible scenarios; one I will define as pessimistic, the other optimistic.

    1) <u>Pessimistic</u>: LEP, HERA, TEVATRON, LHC, will prove the validity of the Standard Model at the highest possible energy and accuracy. In this case subnuclear physics will not survive. Therefore a group is necessary to think about the highest possible energy and the utmost luminosity, now.

    2) <u>Optimistic</u>: LEP, HERA, TEVATRON, LHC, will provide great discoveries. In this case the energy jump would be even more compulsory. But the implementation of a (100 + 100) TeV collider needs at least ten years.

In any case, be the truth on the pessimistic or the optimistic side, we need <u>now</u> to care seriously, i.e. thinking of and implementing R&D projects able to prepare the basis for the highest possible levels of energy and luminosity to become real in a reasonable time.

Some might say that it would be better to wait for new ideas for a new collider to be built. But past experience has shown that to wait is not a wise choice.

In fact, a long time is needed to transform new ideas into reality. Two examples should suffice:

    i) Superconducting high-field magnets, first proposed in 1961, became "reality" in 1986 (the Tevatron): 25 years were required.

    ii) Collective field accelerators were proposed by Veksler, Budker and Fainberg in 1956: a third of a century later, there is still no practical design for a high-energy machine based on these ideas.

In this context it is important to notice that the Eloisatron design is based on extrapolation from known facts and technologies. Nevertheless, the development of

> new acceleration techniques,

> new superconductivity technology (at high temperatures),

should be encouraged for the full-scale project.

For nearly two decades nothing "unexpected" has been discovered, despite the jump in the energy level by an order of magnitude. The list of missing "unexpected" results includes LEP, HERA and the supercolliders at CERN and FERMILAB.

This should be compared with the results obtained [188] from the early fifties on, when totally "unexpected" results were obtained with the advent of i) the proton synchrotrons at CERN and BNL [(C ≠; P ≠; CP ≠; T ≠ and the J particle]; ii) the electron linear accelerator at SLAC (scaling); iii) the (e^+e^-) colliders at ADONE and SLAC (heavy lepton and the J/ψ family of new particles); iv) the 400 GeV proton synchrotron at FERMILAB (Υ states).

Putting together the experimental missing results of these last (nearly two) decades and the theoretical (lack of) understanding of the energy threshold for new physics, there is one and only one wise conclusion today. Our physics needs not a small, but the biggest possible jump in energy and it is therefore very risky to wait and see the results of LHC (< 10% ELN), whose energy could still be too low for the physics beyond the Standard Model to be discovered.

III.10   *Geophysical and Civil Engineering Studies for Cost-Estimates.*

The problem of finding a site for such a large machine as Eloisatron with appropriate geophysical characteristics has been investigated by ING (Istituto Nazionale di Geofisica) in a 1985 study. The results of this study are summarized in Fig. III.10.1. At least seven sites, in Italy, are suitable for the Eloisatron ring.

The civil engineering problems have been investigated in a specialized Working Group. Depending on the flatness of the ground, the machine could be built partly as prefabricated tract (Fig. III.10.2) and partly in a tunnel (Fig. III.2.3). The cost of this solution has been estimated to be no higher than other possible solutions such as the pipe-line, once all needed items are included.

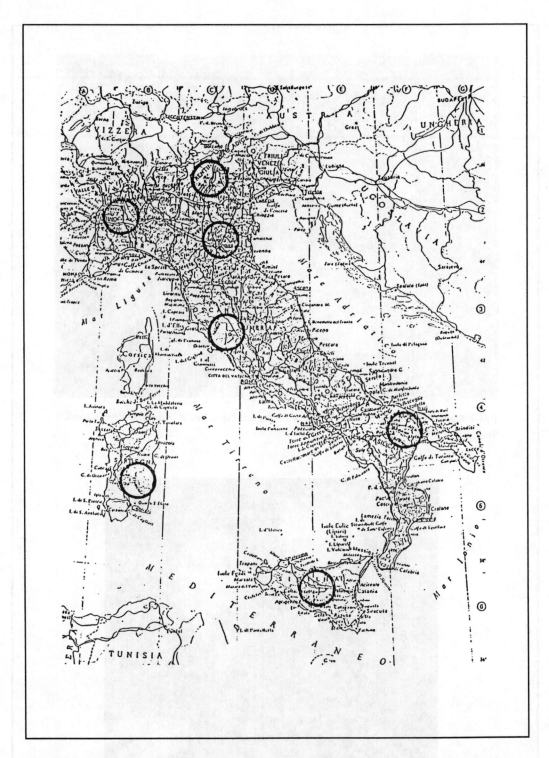

Fig. III.10.1: (Figure from Reference 188). Possible sites for the large Eloisatron ring. [Study made by the Italian National Institute for Geophysics (ING), 1985].

P.M.S. Blackett's Statement and the Director of the Blackett Laboratory at Imperial College in London, Thomas W.B. Kibble.

Erik Rudberg and the author, Stockholm, 1976.

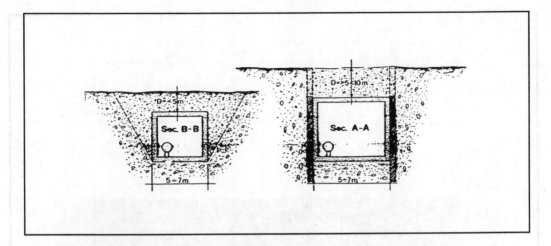

Fig. III.10.2: (Figure from Reference 188). The typical cross-sections for prefabricated tracts.

IV – CONCLUSIONS: *Blackett and Russell (1955-1997)*.

The best conclusion I can think of is to mention an interesting dinner I had in Blackett's house with Mrs Blackett (a charming genuine Florentine lady) and Blackett's great friend, Bertrand Russell. The purpose of the dinner was to discuss the exact formulation of the principle of relativity by Galileo Galilei since Russell had written a book on Einstein's relativity. At some point the conversation switched to physics and CERN.

Russell said[*], «You physicists believe that the governments support your researches because you are the highest intelligence capable of discovering the truth. Well no. The reason why you are so successful in getting a lot of financial support is because your research work gives rise to more and more sophisticated war technologies. You physicists, in order to perform an experiment, would sell your soul to the devil.» And looking at me, he added, «Have you ever thought why my friend has succeeded to convince the British government to finance the new laboratory where you are working? You think that it is because you are searching for the truth. Listen please, you are very young[**]. If one day the East-West confrontation ends, you had better change activity. The governments will stop supporting your work.»

Both Blackett and Russell were convinced that the East-West confrontation was unavoidable and were both engaged in trying to postpone it as much as possible. Some

---

(*)   The quotation is based on my recollection of the evening's discussion. The words may have been different but their meaning is exact [213].

(**)   1955.

The author, Samuel C.C. Ting, Luc Montagnier, H.H. John Paul II, Kai M.S. Siegbahn, Tsung Dao Lee, Edward Teller at Erice (8 May 1993).

days later Blackett told me not to worry about Russell's views. "He does not like us physicists because Kurt Gödel was a physicist who later became engaged in mathematical logic". In 1931 Kurt Gödel discovered a theorem [214], which is probably the most important in all mathematics. The theorem destroyed the Principia Mathematica of Russell and Whitehead.

It is now more than four decades later and what has happened with SSC and LHC seems to be exactly in line with Bertrand Russell's prediction. It is vital that we, subnuclear physicists, realize the importance of making everybody understand the intrinsic value of our research work and the outstanding contributions it makes to peace technology. If this had happened earlier, SSC would not have been cancelled nor would LHC be confronted with the difficulties that continuously delay its implementation.

Bertrand Russell was the interpreter of the real impact of our physics on the culture of our time. Neither the media nor the tax-payers understand either the *cultural value* of our research work, or the fact that very important *technological developments* are generated in subnuclear physics. We must arouse the public's interest and gain its support. Everybody has heard of the Big-Bang, whereas the Standard Model — that great synthesis resulting from 400 years of Galilean Science — is practically unknown to the general public.

The immensely rich area of knowledge connecting the pre-Big-Bang and the Standard Model is yet to be explored.

We subnuclear physicists have the privilege and the responsibility of ensuring that this field of advanced scientific research continues to progress and does not die out adiabatically.

V – ACKNOWLEDGEMENTS.

The Galvani Bicentenary Celebrations were related to three other events:

    i)   the centenary of the electron discovery by J.J. Thomson;

    ii) the centenary of the Italian Physical Society;

    iii) the 50th anniversary of the birth of Subnuclear Physics.

The Academy of Sciences of the oldest University in the world started the Bicentenary Celebrations of Luigi Galvani with the first event being the opening lecture of the Academic year on March 7, 1998. The other events were: i) The 50th anniversary of Subnuclear Physics: Erice, September 2, 1997 and INFN-Gran Sasso Lab: L'Aquila, May 8, 1998; ii) The centenary of the Italian Physical Society: Como, October 28, 1997; iii) The centenary of the electron discovery: INFN-Frascati National Lab: November 12, 1997. This review is an extended version of these lectures.

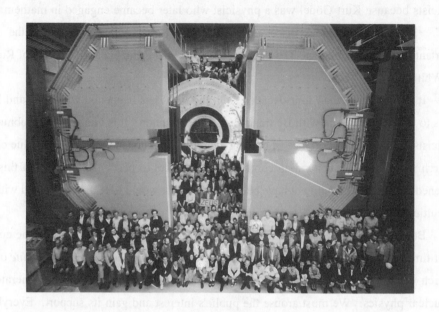

The ZEUS Group.

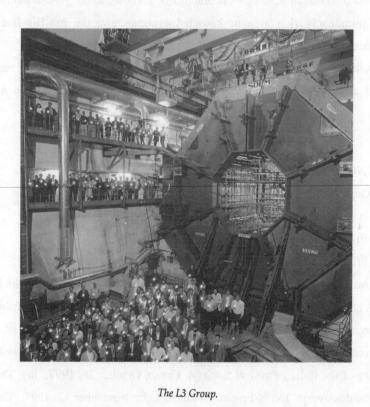

The L3 Group.

I am grateful to the President (O. Barnabei) of the Academy of Sciences, to the Chairman (P. Pupillo) of the Science Faculty and to the Rector (F. Roversi Monaco), of the University of Bologna, to the President (L. Maiani) of the Italian National Institute for Nuclear and Subnuclear Physics (INFN) and to the President (R.A. Ricci) of the Italian Physical Society (SIF) for their friendly insistence to publish a written and expanded version of my lectures. I enjoyed writing this review, on the 50th Anniversary of Subnuclear Physics, because it allowed me to recall some of the most exciting events in my scientific life.

I would also like to thank for their advice and contributions all members of the 200 TeV Club, especially T.D. Lee, G. 't Hooft, G. Veneziano, and all my collaborators in the ELN project, physicists, engineers, technicians whose names are reported in the following pages:

K.S. Adhao, Z. Aftab, K.B. Agiawe, C. Aglietta, P.B. Agrawal, Z. Ahsan, J. Alberty, M. Ali, B. Alpat, J. Alsford, C. Alvisi, E.D. Alyea, G. Ambrosi, Q. An, F. Anghinolfi, F. Anselmo, P. Antonioli, D. Antreasyan, G. Anzivino, A. Arefiev, M. Arneodo, R. Arnold, F. Arzarello, A.V. Asole, P. Aspell, B.H. Aurade, P.V.K.S. Baba, V.B. Badhe, G. Badino, S.D. Bagchi, M.K. Bagde, K.R. Bagree, D.R. Bahekar, F.N. Bai, J.Z. Bai, U.D.N. Bajpai, J.S. Bakare, C. Baldanza, Y. Ban, A.M. Band, G. Bandyopadhyay, S. Bandyopadhyay, H. Banerjee, S.K. Banerjee, A.N. Bannore, G. Barbagli, E. Barberio, G. Bari, T. Barillari, W. Barletta, M. Basile, M. Bassetti, A. Bassi, B. Basu, D. Basu, R. Battiston, C. Baudoin-Bijst, A. Bazzani, U. Becker, S.G. Bedi, S.Y. Bedore, L. Bellagamba, R.B. Belorkar, N.G. Belsare, M. Benot, P. Benvenuto, J. Berbiers, J. Berdugo, V.S. Berezinsky, L. Bergamasco, F. Bergsma, A. Bertin, R. Bertin, V.S. Bhagde, D.R. Bhakre, P.D. Bhalerao, K.S. Bhamra, N.C. Bhattacharya, T. Bhattacharya, S.N. Bhongle, S. Bianco, S.S. Bijwe, N. Bingefors, D. Bisello, A. Bizzetti, R.K. Bock, S.M. Borikar, M.S. Borkar, D. Boscherini, M. Bosteels, R. Bouclier, M. Bramhall, E. Braun, G. Bruni, P. Bruni, R. Bruzzese, F. Bugatti, J. Burger, D.K. Burghate, A. Bussolotti, V. Buzuloiu, N. Cabibbo, X.D. Cai, M. Campbell, Y. Cao, L. Caputi, G. Cara Romeo, R. Casaccia, C. Castagnoli, A. Castellina, A. Castelvetri, H. Castro, S. Ceresara, S.M. Chandekar, J.M. Chapuis, G. Charpak, U.K. Chaturvedi, B.F. Chen, C.Q. Chen, H.J. Chen, J.P. Chen, K. Chen, L.M. Chen, P.G. Chen, R. Chen, S. Chen, S.Y. Chen, W.Y. Chen, M. Cherian, E. Chesi, M. Chiarini, J.A. Chinellato, S.P. Chiney, J. Christiansen, E. Christofel, L. Cifarelli, F. Cindolo, G. Cini, F. Ciralli, M. Civita, A. Clare, Z. Coa, E. Colavita, G. Conforto, S. Cong, F. Coninckx, A. Contin, S. Cooper, M. Costa, I. Crotty, H.C. Cui, R.Y. Cui, S.Z. Cui, X.T. Cui, X.Y. Cui, G. D'Ali, C. D'Ambrosio, I. D'Antone, S. D'Auria, V.L. Dadykin, G.L. Dai, C.K. Damle, D.N. Darang, M. Dardo, T.K. Das, M. Dasgupta, B. Datta, B.K. Datta, R. Datta, U.R. Datta, S. Datta Majumdar, D. Davasumony, S. De Pasquale, V. de Sabbata, R. De Salvo, C. Del Papa, S. Deng, Y.P. Deng, S.L. Deoghare, V.S. Deogoankar, D.W. Deshkar, A.A. Deshmukh, B.T. Deshmukh, D.T. Deshmukh, K.C Deshmukh, K.G. Deshmukh, R.M. Deshmukh, V.R. Deshmukh, A.M. Deshpande, G.T. Deshpande, M.V. Deshpande, P.D. Deshpande, A. Desilva, P. Destruel, M. Deutsch, J. deWitt, J.A. Dhanaraj, S.B. Dhannasare, G.B. Dhanokar, R.V. Dharaskar, M.V. Dhargave, M.T. Dharmadhikari, T.N. Dhobe, O. Di Rosa, D. Dorfan, L.G. Dos Santos, M.T. Dova, W.F. Du, X.J. Du, E. Duchovni, J. Dupont, J. Dupraz, M. Dutta, J. Egger, T. Ekelöf, J. Ellis, R.B. Elgira, R.I. Enikeev, C.C. Enz, Y. Ermoline, F.L. Fabbri, J.P. Fabre, C. Fang, S.X. Fang, R.N. Faustov, P. Feraudet, S. Ferrara, L. Ferrario, S. Finelli, F. Fiori, E. Fishbach, D. Fong, P. Ford, F. Frasconi, M. French, M. Fuchs, M. Fukushima, W. Fulgione, K. Gabathuler, P.M. Gadkari, V.J. Gaikwad, Y. Galaktionov, S. Galassini, P. Galeotti, J. Galvez, C. S. Gao, W. Gao, W.X. Gao, M. Gasperini, J. Gaudaen, K.N. Gawande, S.G. Gawande, S.W. Gawande, M.H. Ghate, S.M. Ghatole, P. Ghia,, G.A. Ghike, S.G. Ghosh, S.N. Ghosh, G.T. Gillies, Y. Giomataris, J.P. Girod, P. Giusti, K. Goebel, V.G. Gokhale, V.M. Gokhale, R. Goldoni, N. Gopalkrishnan, Y Gorodkov, A. Gougas, M. Gourdin, F. Grianti,

C. Grinnel, C. Gu, M.P. Gu, S.D. Gu, S.H. Gu, F.Q. Guang, M. Guanziroli, M. Guerzoni,
W.X. Guo, Y. Guo, Z.Y. Guo, V.D. Gupta, H. Güsten, J.L. Guyonnet, T. Gys, E.S. Hafen, Q. Han,
W. Hao, P. Haridas, A. Hasan, D. Hazifotiadu, M.J. He, E. Heijne, T. Henkes, G. Herten,
W.B. Hirurkar, H. Hoorani, M. Hourican, G. Hu, J.L. Hu, J.W. Hu, L.D. Hu, K.X. Huang,
L. Huang, N. Huang, Y.Z. Huang, P.P. Huddar, S.I. Hussain, G. Iacobucci, M. Ikram, M.F. Ingalgi,
A.G. Ingle, B.G. Ingle, R.P. Ingole, N. Inove, M. Italiani, J. Jain, S.K. Jain, R.K. Jambhorka, P. Jarron,
Y.W. Jia, C.H. Jiang, Y.L. Jiang, Q.S. Jin, Y.K Jin, A.S. Joalikar, J.P. Jobez, D.N. Joharapurkar,
K. Johnsen, V.B. Johri, C. Joram, S.V. Joshi, D.K. Kadu, J.M. Kale, M.G. Kale, B. Kaleem Minhas,
Y. Kamyshkov, T.M. Karadei, D.M. Kataria, M. Kaur, S.A.H. Kbatil, D.G. Kelkar, G.M. Kesharwani,
G.S. Khadekar, M.T. Khadse, F.F. Khalchukov, G.A.N. Khan, M.Q. Khan, R.A. Khan, S. Khan,
A.T. Khandare, M.L. Khandelwal, R.B. Kharade, T.M. Kharade, R.B. Kharat, U.U. Kharat,
N.W. Khodragade, S.M. Khodragade, S. Khokhar, V. Khoze, T. Kiltamura, M.S. Kishirsagar,
W. Kluge, E.V. Ko, M.S. Korde, P.V. Kortchaguin, V.B. Kortchaguin, A.P. Kowale, B.P. Kowale,
W. Krisher, F. Krummenacher, B.L. Kuang, V.A. Kudryavtsev, A.B. Kulkarni, D.K. Kulkarni,
R.S. Kulkarni, V. Kumar, G.N. Kumbhare, A. Kuzucu, G. La Commare, I. Laakso, J.C. Labbé,
G.R. Labde, S.A. Ladhake, T. Lainis, G. Landi, P.F. Lang, R.B. Lanjewar, H. Larsen, K. Lau,
G. Laurenti, T.D. Lee, A.R. Leo, M. Leo, M. Letheren, H. Leutz, G. Levi, L. Levinson, B.M. Li, C. Li,
D.S. Li, D.Z. Li, F.B. Li, G.L. Li, J. Li, J.C. Li, J.G. Li, M. Li, Y.M. Li, Q. Lin, D.C. Ling, B. Lisowski,
A. Litke, D.K. Liu, J. Liu, S.Y. Liu, T. Liu, X. Liu, Y. Liu, Y.C. Liu, C. Liverani, C. Ljuslin,
G.C. Lokre, L. Lone, S.D. Long, C.G. Lu, J.H. Lu, L. Lu, W.D. Lu, G. Luches, D. Luckey,
K. Luebelsmeyer, H.S. Lunge, Y.X. Luo, Z.H. Luo, J. Ma, J.M. Ma, Z. Ma, G. Maccarrone,
J.K. Madhugiri, R.K.W. Mahalle, S.N. Mahapatra, M.A. Mahure, C.K. Majumdar, M. Malavasi,
A.S. Malguin, R. Malik, B.N. Mandal, S.V. Manmode, J.L. Manputra, M. Mansoor Ali, H.S. Mao,
A. Marchioro, A. Margotti, M. Marino, S. Marmi, T. Massam, C.G. Mathe, T. Matsuda,
T. Matsuura, D. Mattern, B.R. Maurya, B. Mayes, P. Mazzanti, G. Meddeler, C.A. Mehare,
K.H. Meier, V.N. Melnikov, S. Meneghini, R. Meng, N. Mengotti Silva, V.R. Metkar, J.L. Mi,
R.S. Mi, Y. Mi, G. Mikenberg, G. Million, Y. Mir, I. Mirza, A. Misaki, K.D. Mishra, R.B. Mishra,
R.S. Mishra, S.B. Mishra, V.G. Miskin, G.H. Mo, S.G. Modak, A.K. Modi, M. Mohan, G. Mohanty,
S.C. Moholkar, G. Molinari, M.R. Mondardini, A. Montanari, B. Monteleone, C. Morello, K. Morey,
G. Mörk, J. Moromisato, F. Motta, N.E. Moulai, R. Mount, A. Mukherjee, S. Mukherjee, K.V. Muley,
V.D. Muley, A.M. Mundhada, B. Musso, S. Muthiah, A.B. Nadange, P.T. Naiekar, S.Y. Nandanwar,
R. Nania, R.Y. Narkhede, S.G. Nasare, V. Nassisi, L. Natrajan, G. Navarra, C. Nemoz,
K.A.N. Nerkar, S. Newett, H. Newman, M.A. Niaz, B.R. Nikhade, F.M. Nirwan, V. O'Shea,
A. Oliva, A. Olsen, N. Ozdes, R.S. Pagrut, J.P. Pal, M. Pal, E. Pallante, F. Palmonari, H.B. Pan,
L. Panaro, N.K. Pande, S.N. Pandey, T.N. Pandey, U.S. Pandey, J.B. Pang, K.G. Pangarkar,
G. Papini, S.C. Paranjape, Y.R. Paranjape, D.V. Parwate, G. Passardi, G. Passotti, C.B. Patil,
M.B. Patil, P.D. Patil, S.P. Patil, B.A. Patki, A. Paul, K.V. Pawar, S.P. Pawar, R. Peccei, P.G. Pelfer,
A.W. Pendharkar, M. Pereira, L. Periale, C. Peroni, E. Perotto, V. Peskov, P. Picchi, D. Piedigrossi,
A. Pieretti, R. Pilastrini, P. Pinna, L. Pinsky, D. Pitzl, V. Pjidaev, I.A. Pless, V. Plyaskin, G. Pocci,
R.B. Pode, H.M. Poharkar, O. Polgrossi, C.N. Potdar, N.B. Potdar, V.K. Pratapwar, G. Prisco,
M. Pu, D. Puertolas, M. Puglisi, H.S. Pundkar, J. Pyrlik, S. Qian, Z.M. Qian, J.M. Qiao, J. Qin,
H. Qiu, N. Qureshi, A. Racz, V.D. Raghatate, S.B. Raichur, V.B. Rajurkar, M. Rama, A.D. Rangari,
S.P. Rant, V.A. Ratate, G.D. Rathod, A.D. Raut, D.R. Raut, W.B. Ren, Z.L. Ren, M. Ricci,
G. Rinaldi, F. Rivera, H.A. Rizvi, C. Rizzuto, P. Rotelli, P. Roumeliotis, R. Roy, T.Z. Ruan, T. Ruf,
M.V.G. Ryassny, O. Ryazhskaya, H. Rykaczewski, H. Rykaszewski, O. Saavedra, N. Sacchetti,
H. Sadrozinski, E. Saletan, R. Salgne, D.L. Samudralwar, D. Sanders, A. Sandoval, L. Sandri,
J.C. Santiard, S. Santini, G. Sartorelli, P. Sartori, S. Sarwar, N.N. Saste, R.V. Satpute, F. Sauli,
W. Scandale, M. Schioppa, J. Schipper, H. Schönbacher, D. Scigocki, M. Scioni, R. Scrimaglio,

J. Seguinot, R. Sehgal, A. Seiden, W. Seidl, J.M. Seixas, S.H. Selukar, D. Sen, M. Sen Gupta, N.D. Sen Gupta, H. Sens, S. Serra, G. Servizi, H.Y. Sha, D.I. Shahare, V. Sharan, A. Sharma, P. Sharp, W.R. Sheldon, V.P. Shelest, B.H. Shen, J.P. Shen, T.J. Shen, V. Shevchenko, Y.S. Shi, V.S. Shiramwar, S.P. Shrinkahnde, W.S. Shu, Y.D. Shu, J.B. Shukla, E. Shumilov, S. Siboni, A. Sigrist, G. Simonet, M.D. Singh, R.D. Singh, S.K. Singh, T. Singh, J. Singh Rana, K.P. Sinha, K. Smith, P.V. Soadekar, V.S. Soitkar, G. Soliani, Ke Songbai, D.N. Soni, M. Spadoni, E. Spencer, L. Sportelli, R.S.L. Srivastav, A. Staiano, M. Steuer, G. Sultanov, L. Sun, S.L. Sun, Y.L. Sun, S.N. Supe, G. Susinno, G.C. Susinno, J. Swain, S.P. Swami, A.A. Syed, V.A. Tabhane, S. Tailhardat, N. Takahashi, Y. Takano, V.K. Tale, V.P. Talochkin, R.K. Talwekar, T.M. Tamhane, H. Tanaka, B.P. Tang, C. Tang, H. Tang, J. Tang, S.M. Tang, M.S. Tapi, E. Tarkowski, M. Taufer, M. Tavlet, S. Tazzari, M.T. Teli, T.J. Telranhe, B.R. Tembhurne, K.J. Teng, A.E. Terraneo, S.C. Thaker, V.A. Thakhare, A.W. Thakre, R.V. Thakre, V.M. Thatte, F. Tian, W. Tian, S.D. Tikar, R. Timellini, S.C.C. Ting, J. Tischhauser, B.P. Tiwari, S.A. Tiwari, J. Tocqueville, G. Torelli, G.C. Trinchero V.S. Tumram, G. Turchetti, A. Turtelli, D.U. Umak, I. Uman, V. Valencic, S. Valenti, P. Vallania, L. Vallini, G. Vanstraelen, G. Venturi, S. Vernetto, I. Vetliski, F. Vetrano, K. Vidya Gaikwad, P. Vikas, U. Vikas, F. Villa, A. Vishnoi, A. Vitale, M. Vivargent, E. von Goeler, L. Votano, T. Wada, M. Wadhwa, D.G. Wadke, S.D. Wadke, R.V. Waghamare, W.U. Waghmode, S.H. Walkey, A.B. Walkhade, B.S. Wang, D.C. Wang, F. Wang, H.F. Wang, H.J. Wang, J. Wang, L.L. Wang, L.Z. Wang, M. Wang, R. Wang, S. Wang, S.H. Wang, S.Q. Wang, T.J. Wang, X.P. Wang, Y.Y. Wang, Z. Wang, P.B. Wani, A.M. Wankhede, P.C. Wankhede, T.V. Warhekar, T.S. Wasnik, Cao Wei, K.Y. Wei, Z.Z. Wei, R. Weinstein, T. Wenaus, H. Wenninger, C. Werner, M. White, M. Widgoff, C. Williams, M. Willutzky, B.Z. Wu, C.C. Wu, M.A. Wu, S.X. Wu, W.M. Wu, W.T. Wu, Y.Z. Wu, Z.D. Wu, B. Wyslouch, B.X. Xi, J.W. Xi, L.G. Xiao, J.L. Xie, P.P. Xie, X.X. Xie, Y.Y. Xie, B. Xu, C. Xu, J.M. Xu, L. Xu, S.W Xu, X..K. Xu, Z.X Xu, J.X Xue, S.T. Xue, R. Yagannathan, V.F. Yakushev, I. Yamamoto, B.H. Yan, J. Yan, T.X. Yan, W.G. Yan, D.J. Yang, G. Yang, Y. Yang, X.G. Yao, C.H. Ye, Q.H. Ye, S.H. Yeh, A.K. Yelne, S.V. Yenkar, G. Yi, Z.S. Yin, Cai Yixing, J.M. You, K. You, T. Ypsilantis, Q.C. Yu, Q.F. Yu, Z.Q. Yu, Z.Z. Yu, W. Yue, N. Yunus, T.M. Zachariah, N. Zaganidis, A. Zallo, G. Zanetti, G. Zatsepin, M. Zeng, B.C. Zhang, C. Zhang, H.S. Zhang, Y. Zhang, Y.P. Zhang, Z.P. Zhang, B. Zhao, D. Zhao, F.G. Zhao, J.B. Zhao, J.J. Zhao, W.R. Zhao, Y.J. Zhao, G.Q. Zheng, L.S. Zheng, S.C. Zheng, Y.C. Zheng, Z.P. Zheng, Cai Zhiguo, S.C. Zhong, J.K. Zhou, S. Zhou, X. Zhou, X.G. Zhou, Y.H. Zhou, F.Q. Zhu, L.S. Zhu, R.Q. Zhu, R.Y. Zhu, S.G. Zhu, X. Zhu, Y.C. Zhu, B. Zhuan, A. Zucchini, M. Zuffa

VI – REFERENCES AND NOTES.

[1] *Fine Structure of the Hydrogen Atom by a Microwave Method*
W.E. Lamb and R.C. Retherford
Phys. Rev. 72, 241 (1947).

[2] *Processes Involving Charged Mesons*
C.M.G. Lattes, H. Muirhead, G.P.S. Occhialini and C.F. Powell
Nature 159, 694 (1947);
and
Observations on the Tracks of Slow Mesons in Photographic Emulsions
C.M.G. Lattes, G.P.S. Occhialini and C.F. Powell
Nature 160, 454 (1947).

[3] *Evidence for the Existence of New Unstable Elementary Particles*
G.D. Rochester and C.C. Butler, *Nature* 160, 855 (1947).

[4] The first "radiative" effects (experimentally) discovered are the Lamb-shift and the "anomalous" magnetic moment of the electron. They are examples of two distinct phenomena. One (the Lamb-shift) can be almost fully (95%) accounted for by a non-relativistic calculation (as done by H.A. Bethe who found 1040 MHz: "*The Electromagnetic Shift of Energy Levels*", *Phys. Rev.* 72, 339 (1947)), while the other (the "anomalous" magnetic moment) cannot. The fact that the gyromagnetic ratio of the electron is predicted by the Dirac equation to be $g = 2$ cannot be accounted for by any non-relativistic description. The magnetic behaviour of an electron in a magnetic field corresponds to the effect caused by the emission and reabsorption of virtual photons on its magnetic moment. The "anomalous" magnetic moment of the electron, i.e. its g value being different from 2, needs a relativistic description of the virtual electromagnetic processes. This was done by J. Schwinger a few months after the experimental discovery of the Lamb-shift: *Phys. Rev.* 73, 416 (1948). In this paper the "anomalous" magnetic moment of the electron was theoretically found to be the by now famous $(\alpha/2\pi)$.

Furthermore, while the Lamb-shift affects only the hydrogen atom, the deviation from QED of an intrinsic property such as the "magnetic" moment of an elementary particle (the electron) is expected to affect other particles as well. In fact — when the muon came in — the "anomalous" magnetic moment of this unexpected particle was considered the crucial check in order to verify its intrinsic properties (see § II.2-2).

[5] *Dal Gran Sasso al Supermondo*
L. Maiani and A. Zichichi, INFN/AE-98/19, July 1998.

[6ᵃ] *The Gran Sasso Project*
A. Zichichi, Proceedings of the GUD-Workshop on "*Physics and Astrophysics with a Multikiloton Modular Underground Track Detector*", Rome, Italy, 29-31 October 1981, INFN, Frascati, 141 (1982).

[6b] *The Gran Sasso Project*
A. Zichichi, INFN/AE-82/1, 28 February 1982.

[6c] *The Gran Sasso Project*
A. Zichichi, Proceedings of the Workshop on *"Science Underground"*, Los Alamos, NM, USA, 27 September-1 October 1982, AIP, New York, 52, (1983).

[7] *The Gran Sasso Laboratory*
A. Zichichi, Proceedings of the International Colloquium on *"Matter Non-Conservation"* - ICOMAN '83, Frascati, Italy, 17-21 January 1983, INFN, Frascati, 3 (1983).

[8] *The Gran Sasso Laboratory and the Eloisatron Project*
A. Zichichi in "Old and New Forces of Nature", Erice 1985, A. Zichichi (ed) Plenum Press, New York and London, 335 (1988).

[9] *Perspectives of Underground Physics: The Gran Sasso Project*
A. Zichichi, Invited Plenary Lecture at the Symposium on *"Present Trends, Concepts and Instruments of Particle Physics"*, in honour of Marcello Conversi's 70th birthday, Rome, Italy, 3-4 November 1987, G. Baroni, L. Maiani and G. Salvini (eds), Conference Proceedings, SIF, Bologna, Vol. 15, 107 (1988).

[10] *The HERA Collider*
V. Soergel, P. Waloschek and B. Wiik, DESY, January 1994.

[11] *The Thin Superconducting Solenoid for ZEUS*
M. Morpurgo, INFN/CERN report 1985;

The Thin Superconducting Solenoid for ZEUS
Q. Lin and M. Morpurgo, INFN/CERN report 1988;

Design Study of a ϕ 19.5 × 36 M Superconducting Solenoid
P. Bruni, S. Ceresara, Y. Li, Q. Lin, B. Musso and A. Zichichi, Proceedings of the *"Applied Superconductivity"* Conference - ASC 1990, Snowmass, CO, USA, 24-26 September 1990, *IEEE Transactions on Magnetics* 27 n. 2, 1969 (1991).

[12] *Alpi project*
P. Dalpiaz, LNF-INFN-001/87;

The ALPI Project at the Laboratori Nazionali di Legnaro
G. Fortuna, R. Pengo, G. Bassato, I. Ben-Zvi, J.D. Larson, J.S. Sokolowski, L. Badan, A. Battistella, G. Bisoffi, G. Buso, M. Cavenago, F. Cervellera, A. Dainelli, A. Facco, P. Favaron, A. Lombardi, S. Marigo, M.F. Moisio, V. Palmieri, A.M. Porcellato, K. Rudolph, R. Preciso and B. Tiveron, *NIM* A287, 253 (1990);

Commissioning of the ALPI post-accelerator
A. Dainelli, G. Bassato, A. Battistella, M. Bellato, A. Beltramin, L. Bertazzo, G. Bezzon, G. Bisoffi, L. Boscagli, S. Canella, D. Carlucci, F. Cervellera, F. Chiurlotto, T. Contran, M. De Lazzari, A. Facco, P. Favaron, G. Fortuna, S. Gustafsson, M. Lollo, A. Lombardi, S. Marigo, M.F. Moisio, V. Palmieri, R. Pengo, A. Pisent, M. Poggi, F. Poletto, A.M. Porcellato and L. Ziomi, *NIM* A382, 100 (1996);

Status of ALPI and Related Developments of Superconducting Structures
G. Fortuna, G. Bisoffi, A. Facco, A. Lombardi, V. Palmieri, A. Pisent and A.M. Porcellato, *Proceedings Linac 96 Conference*, C. Hill and M. Vretenar (eds), CERN 96-07, 905 (1996).

[13] *Commissioning of the K800 INFN Cyclotron*
L. Calabretta, G. Cuttone, S. Gammino, P. Gmaj, E. Migneco, G. Raia, D. Rifuggiato, A. Rovelli, J. Sura, A. Amato, G. Attinà, M. Cafici, A. Caruso, G. De Luca, S. Pace, S. Passarello, S. Pulvirenti, G. Sarta, M. Sedita, A. Spartà, F. Speziale, E. Acerbi, F. Alessandria, G. Bellomo, C. Birattari, A. Bosotti, C. De Martinis, E. Fabrici, D. Giove, P. Michelato, C. Pagani, L. Rossi, G. Baccaglioni, W. Giussani and G. Varisco, in *Proceedings XIV Conference on Cyclotrons and their Applications*, Capetown, 1995, J.C. Cornell (ed), World Scientific, 12 (1996);

First Operations of the LNS Heavy Ions Facility
L. Calabretta, G. Ciavola, G. Cuttone, S. Gammino, P. Gmaj, E. Migneco, G. Raia, D. Rifuggiato, A. Rovelli, J. Sura, V. Scuderi, E. Acerbi, F. Alessandria, G. Bellomo, A. Bosotti, C. De Martinis, D. Giove, P. Michelato, C. Pagani and L. Rossi, *NIM* A382, 140 (1996).

[14] *The INFN-LASA Lab*
Acerbi et al., *Proceedings 9th International Conference on Cyclotrons and their Applications*, Caen (France), Ed. de Physique, 169 (1981).

[15] *Reports from LEP*
U.F. Becker, in "*From the Planck Length to the Hubble Radius*", Erice 1998, World Scientific, to be published.

[16] The progress of Subnuclear Physics as reported in the Erice School books whose titles follow. See also Appendix.
1) *Strong, Electromagnetic and Weak Interactions* (1963); 2) *Symmetries in Elementary Particle Physics* (1964); 3) *Recent Developments in Particle Symmetries* (1965); 4) *Strong and Weak Interactions - Present Problems* (1966); 5) *Hadrons and their Interactions* (1967); 6) *Theory and Phenomenology in Particle Physics* (1968); 7) *Subnuclear Phenomena* (1969); 8) *Elementary Processes at High Energy* (1970); 9) *Properties of the Fundamental Interactions* (1971); 10) *Highlights in Particle Physics* (1972); 11) *Laws of Hadronic Matter* (1973); 12) *Lepton and Hadron Structure* (1974); 13) *New Phenomena in Subnuclear Physics* (1975); 14) *Understanding the Fundamental Constituents of Matter* (1976); 15) *The Whys of Subnuclear Physics* (1977); 16) *The New Aspects of Subnuclear Physics* (1978); 17) *Pointlike Structures Inside and Outside Hadrons* (1979); 18) *The High-Energy Limits* (1980); 19) *The Unity of Fundamental Interactions* (1981); 20) *Gauge Interactions: Theory and Experiments* (1982); 21) *How far are we from the Gauge Forces* (1983); 22) *Quarks, Leptons and their Constituents* (1984); 23) *Old and New Forces of Nature* (1985); 24) *The Super World I* (1986); 25) *The Super World II* (1987); 26) *The Super World III* (1988); 27) *The Challenging Questions* (1989); 28) *Physics up to 200 TeV* (1990); 29) *Physics at the Highest Energy and Luminosity: To Understand the Origin of Mass* (1991); 30) *From Superstrings to the Real Superworld* (1992); 31) *From Supersymmetry to the Origin of Space-Time*

(1993); 32) *From Superstring to Present-day Physics* (1994); 33) *Vacuum and Vacua: the Physics of Nothing* (1995); 34) *Effective Theories and Fundamental Interactions* (1996); 35) *Highlights: 50 Years Later* (1997).

Vol. 1 was published by W.A. Benjamin, Inc., New York; 2–8 and 11–12 by Academic Press, New York and London; 9–10 by Editrice Compositori, Bologna; 13–29 by Plenum Press, New York and London; 30–36 by World Scientific.

[17]   *A Proposal to Search for Leptonic Quarks and Heavy Leptons Produced by ADONE*
M. Bernardini, D. Bollini, E. Fiorentino, F. Mainardi, T. Massam, L. Monari, F. Palmonari and A. Zichichi, *INFN/AE*-67/3, 20 March 1967; see also

Limits on the Electromagnetic Production of Heavy Leptons
V. Alles-Borelli, M. Bernardini, D. Bollini, P.L. Brunini, T. Massam, L. Monari, F. Palmonari and A. Zichichi, *Lettere al Nuovo Cimento* 4, 1156 (1970);

Limits on the Mass of Heavy Leptons
M. Bernardini, D. Bollini, P.L. Brunini, E. Fiorentino, T. Massam, L. Monari, F. Palmonari, F. Rimondi and A. Zichichi, *Nuovo Cimento* 17A, 383 (1973); and Ref. 98.

[18]   Kaluza and Klein were the first [19, 20] to think and propose that the electromagnetic forces could be described in terms of an extra dimension of space to be added to the standard Lorentz space-time as the one illustrated in Fig. II.1.2. Thus, instead of $(3 + 1)$ dimensions, the Lorentz space should have $(4 + 1)$ dimensions. The extra space dimension, the 5th one, would be compactified around a circle, thus producing the $U(1)$ symmetry which generates the electromagnetic forces. The latest news is that our world has its origin in 11 bosonic dimensions (10 space + 1 time) and 32 fermionic dimensions. In Reference 21 we shall try to summarize how we go from these 43 dimensions of the superspace to our world with $(3 + 1)$ bosonic dimensions and the gauge forces $SU(3) \times SU(2) \times U(1)$ plus 3 families. Note that the time dimension is always one. The space dimensions are 10. In $(10 + 1)$ dimensions supergravity has no gauge group, i.e. no vector fields. If one dimension is compactified, this becomes a coupling. In fact superstring theory in $(9 + 1)$ dimensions plus a coupling is equivalent to supergravity in $(10 + 1)$ dimensions, as shown by P. Horava and E. Witten [22]. Thus, the origin of the fictitious spaces with one, two and three complex dimensions is in the 43 dimensions of the superspace [see Reference 21 for a discussion of these topics].

[19]   *Zum Unitätsproblem der Physik*
T. Kaluza, *Sitz. Phreuss Akad. Wiss.* K11, 466 (1921).

[20]   *Quantentheorie und fünfdimensionale Relativitätstheorie*
O. Klein, *Z. Phys.* 37, 895 (1926).

[21]   *From Supersymmetry to Superstring*
We have emphasized that the unification of the gauge couplings needs SuperSymmetry: i.e. bosons and fermions on equal footing. This symmetry is also needed to keep the Fermi and the Planck scales separated (hierarchy). The amount of work in this field of theoretical speculations has been really impressive,

during the last two decades. The sequence is as follows: SuperSymmetry and SuperGravity, String theory, SuperString theory. In what follows we try to report all that in a telegraphic way. At the end we discuss the physics background of SuperGravity and SuperString.

Supersymmetry and Supergravity

Symmetry Operators in order to be of value for physics must not change with time: i.e. they must commute with the Hamiltonian.

The existence of Symmetry Operators which commute with the Hamiltonian and lead to a different spin was not known. That such Symmetry Operators could exist was a big discovery. This is the first step of SuperSymmetry.

The next step was to make the Symmetry Operators local, i.e. to be free to make different transformations at different space-time points. This freedom generates the gauge bosons of SuperSymmetry, which include the graviton, i.e. a gravitational field coupled à la Einstein to a tensor field, and another particle called the gravitino (with spin = 3/2). As any half integer field, the gravitino cannot carry a long range force. This is how SuperGravity started. And it goes on since the graviton comes in with a set of particles which includes not only the gravitino, but also a spin 1 particle, called the graviphoton.

The graviphoton attracted a lot of interest because it would generate what became to be known as antigravity. In fact, if a vector particle were gravitationally coupled, then we should expect to have not only attractive gravitational forces (as it is for the tensor field) but also repulsive gravitational forces. Experimental search for the existence of antigravity has failed — so far — to prove the existence of such a vector force in gravitation.

Note that SuperGravity is a theory which has a point-like mathematical formalism and puts bosons and fermions on equal footing. However SuperGravity has solitons as solutions of the equation of motion. Solitons are extended objects. This is how non point-like structures enter in SuperGravity. And this is how SuperGravity is linked to SuperString theory. Another relevant point is that, in (10 space + 1 time) dimensions SuperGravity has no vector field, i.e. gauge forces as those needed for the standard model. If one dimension is compactified, this becomes a coupling. P. Horava and E. Witten [22] have shown that SuperGravity in (10 + 1) dimensions is equivalent to SuperString theory in (9 + 1) dimensions plus a coupling.

String Theory

The basic idea of the string theory is that, instead of the concept of a point particle, one has an extended object. A string can be open with two ends, or closed with no ends. The most recent developments require compactification on a segment, thus not on a circle, closed string (see later).

An important feature of a string is its tension which has the dimension of $(\ell)^{-2}$. Moreover, the string once it is stretched can vibrate and it can rotate. The conclusion is that a string has a lot of excitations, which cannot escape quantization. These excitations correspond to particles of different masses. Thus, string theory gives an infinite spectrum of particle masses, but has only one parameter: the string tension. Since the gravitational constant is also dimensional and has to come out of the string theory, it is necessary that the string tension and the gravitational constant are closely related.

Superstring Theory and Supergravity

This is an extension of the string theory, which incorporates fermionic degrees of freedom. SuperString is a theory which starts with the mathematical structure of extended (non point-like) objects and puts bosons and fermions on equal footing, of course. SuperString has automatically a scale: the tension of the extended object, but needs a dimensionless gauge coupling. It was immediately realized that the SuperString works at most with 10 bosonic dimensions (9 space + 1 time). This is for two reasons. One is because we want no massless particles with spin higher than $2\hbar$, since the highest massless spin particle known is the graviton. The second reason is because we need a scalar field. Scalar, in order to be sure that, whatever this field does, its effects will be Lorentz invariant. In eleven dimensions there is no massless scalar field. The scalar field is needed in order to provide the dimensionless gauge coupling in the string theory. This gauge coupling is in fact the vacuum expectation value (vev) of the scalar field (called the dilaton). The fact that the vev of the dilaton must be $\neq$ zero is a very important result of string theory. The vev is undetermined, but the crucial point is that it must be $\neq$ zero, otherwise we would have no string gauge coupling.

On the other hand, a string being an extended object, it provides a fundamental length, which must be linked to the Planck length. In addition, string theory must have a dimensionless gauge coupling, otherwise we would have non interacting strings. As discussed above, the SuperString gauge coupling corresponds to the vacuum expectation value of the dilaton field. It is this gauge coupling which is connected with the unified gauge couplings of the Standard Model ($\alpha_1\ \alpha_2\ \alpha_3$). SuperString theory does not provide the fundamental length. This is derived from the Planck length. SuperString theory does not provide the vev of the dilaton. This is taken to be equal to α_{GUT} at the E_{GUT} scale. Since the only scale is the Planck scale, the particle masses are either zero or multiple of the Planck mass. The ordinary particles have to be in the mass-less sector. At the end we want the Standard Model to come out. If the gravitational field has to come out as well, one must hope for a spontaneous compactification in which 6 of the 10 dimensions curl up into small spaces, leaving four bosonic dimensions for the real space and time of our world. The problem arises: how the compactification has to be, in a circle (à la Kaluza-Klein) or in a segment?

The new development is that if we start with SuperGravity in (10 + 1) dimensions and compactify on a segment (not on a circle à la Kaluza-Klein), the end points of the segment each have 10 dimensions. This is SuperString theory in (9 + 1) dimensions plus a coupling as proved by Horava and Witten [22], quoted earlier. Imposing anomaly cancellation, gauge groups appear and these are $E_8 \times E_8$. It is the compactification into a segment and the condition of anomaly cancellation which produce the gauge groups $E_8 \times E_8$. The compactification into a segment reduces the number of fermionic dimensions from 32 to 16. If the compactification were on a circle (à la Kaluza-Klein), the number of fermionic dimensions would remain 32. Another important detail: the length of the segment gives the coupling. A long segment corresponds to a strong coupling; a short segment corresponds to a weak coupling for the SuperString theory. The last step is the compactification of the 6 remaining bosonic dimensions in a Calabi-Yau manifold [23, 24]. This manifold has the property of reducing the 16 fermionic dimensions into 4 fermionic dimensions (i.e. N = 1 supersymmetry) and giving rise to the gauge groups $E_6 \times E_8$. The E_6

group contains $SU(3) \times SU(2) \times U(1)$ and this is the Standard Model. The group E_8 remains hidden and all particles of our world are singlets. The particles of E_8 communicate with the particles of the Standard Model via gravitational interactions. The hope is that this hidden E_8 sector, gravitationally coupled to the Standard Model, is going to be the source of supersymmetry breaking [25]. An important detail is that, in general, the 6-dimensional Calabi-Yau manifolds have the property that right and left handed states are not equal in number. Therefore parity violation is in the structure of space. This development is called Heterotic SuperString theory. One of the important points in the development of the Heterotic SuperString theory is that the number of families can be derived [26] from the topological properties of the Calabi-Yau manifold [27]. In other words, «why are there three families» has as answer another question: why that particular topological property in the Calabi-Yau manifold? The final result is that the sector of very light masses (compared to the Planck mass) is described by N = 1 SuperSymmetry, which needs, of course, to be broken. How? Hopefully, as mentioned above, by the heavy particle sector which is a sort of "shadow matter" which interacts with our world, described by the Standard Model, only via gravitation.

Another hope is to develop the SuperString theory not in a flat but in a curved space so as to look like Einstein theory from the very beginning. Once compactification is active, you find again particles which are very massive (compared to the Planck mass) and mass-less. These last ones must acquire small masses via some mechanisms, as for example, spontaneous symmetry breaking.

Supergravity and Superstring: the Physics Background

The physics of SuperGravity can be traced back to the observation that "bosons" and "fermions" are gravitationally coupled in the same way. Gravitation does not distinguish bosons from fermions. Matter made out of bosons and matter made out of fermions cannot be distinguished if we switch on gravity. On the other hand, matter is made of fermions but can exhibit "bosonic" properties such as "superfluidity" and "superconductivity". In this last case, pairs of fermions (two electrons = the Cooper pairs) behave as bosons and electromagnetism exhibits the bosonic properties which are absent in standard QED: i.e. when spinors are coupled to the gauge boson of QED not coupled among themselves to build up bosonic states.

In a sense, SuperGravity should have been discovered earlier since it is the counterpart of Superfluidity and of Superconductivity. Here matter exhibits "bosonic" properties. In SuperGravity, the gravitational forces exhibit "fermionic" properties.

The physics of SuperString can be traced back to the search for a fundamental length. How to connect the non point-like structure of the mathematical description of natural phenomena to physics is the great challenge. The big steps are, first, the discovery that the String-tension is a dimensional-full coupling and can therefore be connected with the Newton gravitational strength, i.e. m_{pl}^{-2}. The second big step is that, in SuperString theory a gauge coupling comes out as the vev (vacuum expectation value) of a scalar field, the dilaton. Thus, the non point-like description has the gravitational strength and a gauge coupling, in (9 + 1) bosonic dimensions. As discussed above, the most recent discovery is that, compactifying (10 + 1) dimensional SuperGravity on a segment, one obtains SuperString in (9 + 1) dimensions which allow to have N = 1 SuperSymmetry and the Standard Model, SU(3) × SU(2) × U(1), with three families.

SuperGravity and SuperString started from very different physics concepts, but end up being equivalent, once the physics content of these two different approaches are carefully developed.

[22] *Heterotic and Type I String Dynamics from Eleven Dimensions*
P. Horava and E. Witten, *Nucl. Phys.* B460, 506 (1996).

[23] E. Calabi, in *Algebraic Geometry and Topology: A symposium in honour of S. Lefschetz*, Princeton University Press, 78 (1957).

[24] *Calabi's Conjecture and Some New Results in Algebraic Geometry*
S.-T. Yau, *Proc. Natl. Acad. Sci.* 74, 1798 (1977).

[25] *Considerations on the Moduli Space of Calabi-Yau Manifolds*
S. Ferrara, in *The Challenging Questions*, Erice 1989, A. Zichichi (ed), Plenum Press, New York and London, 103 (1990).

[26] *Effective Lagrangians for Superstring Compactification*
S. Ferrara, in *The Superworld III*, Erice 1988, A. Zichichi (ed), Plenum Press, New York and London, 77 (1990).

[27] *Heterotic and Type II Superstrings Compactified on Calabi-Yau Manifolds*
S. Ferrara, in *Physics up to 200 TeV*, Erice 1990, A. Zichichi (ed), Plenum Press, New York and London, 155 (1991).

[28] *Particle Creation by Black Holes*
S.W. Hawking, *Commun. Math. Phys.* 43, 199 (1975).

[29] *Scale Factor Duality for Classical and Quantum Strings*
G. Veneziano, *Phys. Lett.* B265, 287 (1991);

Pre-Big-Bang in String Cosmology
M. Gasperini and G. Veneziano, *Astropart. Phys.* 1, 317 (1993).

For a complete collection of papers on the PBB scenario see *http://www.to.infn.it/~gasperin/*.

[30] The work of Veneziano et al. refers to what happens if we try to go even further backwards in time. The traditional attitude is that the question does not make sense since the universe, and space-time itself, originated from a singularity at t = 0. However, in string theory, we expect the singularity to be absent because of the softening of all interactions above the string scale. Indeed there is no t = 0, but the quantum of time $\Delta t = 10^{-42}$ s. In this case it should be possible to continue our travel backwards in time and for example see what happens after 10^{19} times the quantum of time in the negative time range: i.e. at $- 10^{-23}$ s. What was there before the Big-Bang? (I will rename it the "ex Big-Bang", ex B-B, since it is no longer a singularity).

A possible scenario is as follows. This phase is characterized by a "long" period of "dilaton-driven inflation" during which, according to the equations of string cosmology, space inflates while the unified coupling α_U (i.e. the dilaton $\phi(t)$) undergoes a similar growth. This is shown schematically in the rightmost part of Fig. VI.1. A patch of space-time of the size of 1 Fermi at 10^0 GeV inflates by a factor 10^{12} in about 10^{-23} s thus reaching a size of about 1 mm, at the 10^{18} GeV scale. The expansion rate of the universe, H, grows from the GeV scale at $-t \simeq -10^{-23}$ s to reach string scale values of 10^{18} GeV, in about 10^{-23} s of expansion time. During the same time interval, the coupling α_U, originally tiny $\alpha_U \simeq 10^{-50}$, increases by about 50 orders of magnitude to reach its present value $\alpha_U = \alpha_{SU} = \alpha_{GUT} \simeq \frac{1}{25}$, of course at 10^{18} GeV. In this way the post-ex B-B phase starting at

$$+ t = \Delta t = + 10^{-42} s$$

produces a primordial universe of the desired characteristics: indeed the 1mm patch easily becomes, through the usual expansion, the 10^{28} cm of our visible universe. The old Big-Bang's instant becomes, in this scenario, the moment at which the Hubble expansion rate reached its highest possible value allowed by string theory: 10^{18} GeV. A direct comparison between α_G and α_1 α_2 α_3 is shown in Fig. VI.2, where there is no change of scale in the vertical axes. Note the shrinkage of α_1^{-1} α_2^{-1} α_3^{-1} when compared with the enormous variation of α_G^{-1} during the time after the "quantum of time" (ex B-B) i.e. on the left part of Fig. VI.2. On the right part we have α_U^{-1} which describes the continuation of the unified gauge couplings $(\alpha_1 \; \alpha_2 \; \alpha_3) \equiv \alpha_{GUT} \equiv \alpha_{SU} \equiv \alpha_U = e^{\phi(t)}$ in the pre-Big-Bang phase investigated by Veneziano. During this period the gravitational coupling α_G is given by α_U times the factor

$$\frac{E^2}{E_{SU}^2}$$

in order to account for the peculiar nature of the gravitational forces, which are coupled to the total energy of a particle. This is why α_G, as shown in Fig. VI.2, changes more rapidly than α_U. Note that there is only one point in the energy scale where all couplings coincide $\alpha_{SU} = \alpha_U = \alpha_G \simeq \frac{1}{25}$. This point corresponds to the string unification scale $E_{SU} \simeq 10^{18}$ GeV.

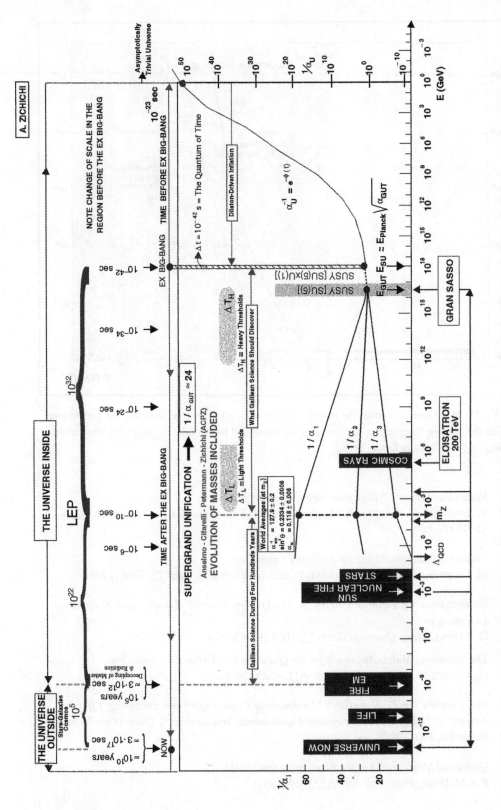

Fig. VI.1: The extension of the physics, illustrated in Fig. II.1.3, to the region before the Big-Bang. In the pre-Big-Bang region there are several possibilities. The one illustrated is just an example due to G. Veneziano et al. The interest in these theoretical speculations [29] is due to their experimental consequences. For example, in explaining the existence of the galactic magnetic fields [32] and in predicting [31] amplitudes, several orders of magnitudes higher than those of the standard inflationary cosmology, for the stochastic background of gravitational waves.

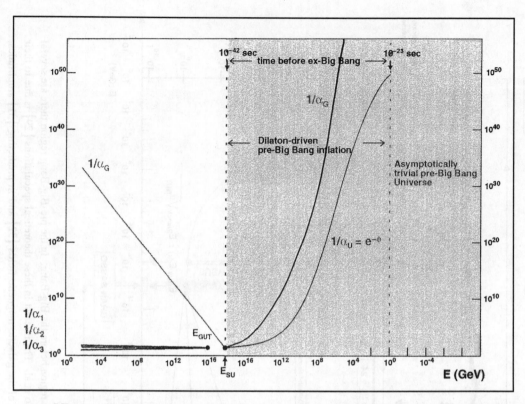

Fig. VI.2: This Figure illustrates the energy dependence of all couplings, including the gravitational one. In the pre-Big-Bang region all couplings are controlled by the dilaton (α_U). The gravitational coupling α_G has the extra factor E^2/E_{SU}^2.

[31] *Relic Gravitational Waves from String Cosmology*
      R. Brustein, M. Gasperini, M. Giovannini and G. Veneziano, *Phys. Lett.* <u>B361</u>, 45 (1995).

[32] *Primordial Magnetic Fields from String Cosmology*
      M. Gasperini, M. Giovannini and G. Veneziano, *Phys. Rev. Lett.* <u>75</u>, 3796 (1995).

[33] *Quantization of Point Particles in (2+1)-Dimensional Gravity and Spacetime Discreteness*
      G. 't Hooft, *Class. Quantum Grav.* <u>13</u>, 1023 (1996); and

      The Scattering Matrix Approach for the Quantum Black Hole: an Overview
      G. 't Hooft, *Int. Journal Mod. Phys.* <u>A11</u>, 4623 (1996).

[34] For a review see G. Veneziano "*An Amusing Cosmology from the String Effective Action*" in "*Effective Theories and Fundamental Interactions*", Erice 1996, World Scientific, 300 (1997).

[35] *Quantised Singularities in the Electromagnetic Field*
      P.A.M. Dirac, *Proc. Roy. Soc.* <u>A133</u>, 60 (1931).

[36]   *The Principles of Quantum Mechanics*
       P.A.M. Dirac, (4th edn), Clarendon Press, Oxford (1958).

[37]   *The Theory of Groups and Quantum Mechanics*
       H. Weyl, Dover Publications, New York (1928).

[38]   *The Positive Electron*
       C.D. Anderson, *Phys. Rev.* 43, 491 (1933);

       Some Photographs of the Tracks of Penetrating Radiation
       P.M.S. Blackett and G.P.S. Occhialini, *Proc. Roy. Soc.* A139, 699 (1933).

[39]   *Theory of Electrons and Positrons*
       P.A.M. Dirac, Nobel Lecture, December 12, 1933.

[40]   *Question of Parity Conservation in Weak Interactions*
       T.D. Lee and C.N. Yang, *Phys. Rev.* 104, 254 (1956).

[41]   *Experimental Test of Parity Conservation in Beta Decay*
       C.S. Wu, E. Ambler, R.W. Hayward, D.D. Hoppes, *Phys. Rev.* 105, 1413 (1957);

       *Observation of the Failure of Conservation of Parity and Charge Conjugation in
       Meson Decays: The Magnetic Moment of the Free Muon*
       R. Garwin, L. Lederman, and M. Weinrich, *Phys. Rev.* 105, 1415 (1957);

       Nuclear Emulsion Evidence for Parity Non-Conservation in the Decay Chain $\pi^+\mu^+e^+$
       J.J. Friedman and V.L. Telegdi, *Phys. Rev.* 105, 1681 (1957).

[42]   *On the Conservation Laws for Weak Interactions*
       L.D. Landau, *Zh. Éksp. Teor. Fiz.* 32, 405 (1957).

[43]   *Remarks on Possible Noninvariance under Time Reversal and Charge Conjugation*
       T.D. Lee, R. Oehme, and C.N. Yang, *Phys. Rev.* 106, 340 (1957).

[44]   *Evidence for the 2π Decay of the K_2^0 Meson*
       J. Christenson, J.W. Cronin, V.L. Fitch, and R. Turlay, *Phys. Rev. Lett.* 13, 138
       (1964).

[45]   *The Analytic S Matrix*
       G.F. Chew, W.A. Benjamin Inc., New York, Amsterdam (1966).

[46]   *Search for the Time-Like Structure of the Proton*
       M. Conversi, T. Massam, Th. Muller and A. Zichichi, *Phys. Lett.* 5, 195 (1963).

[47]   *The Leptonic Annihilation Modes of the Proton-Antiproton System at 6.8 $(GeV/c)^2$
       Timelike Four-Momentum Transfer*
       M. Conversi, T. Massam, Th. Muller and A. Zichichi, *Nuovo Cimento* 40, 690 (1965).

[48]   To the best of my knowledge, the CPT Theorem was first proved by W. Pauli in his
       article "*Exclusion Principle, Lorentz Group and Reflection of Space-Time and*

Charge", in "Niels Bohr and the Development of Physics" [Pergamon Press, London, page 30 (1955)], which in turn is an extension of the work of J. Schwinger [*Phys. Rev.* 82, 914 (1951); *"The Theory of Quantized Fields. II."*, *Phys. Rev.* 91, 713 (1953); *"The Theory of Quantized Fields. III."*, *Phys. Rev.* 91, 728 (1953); *"The Theory of Quantized Fields. VI."*, *Phys. Rev.* 94, 1362 (1954)] and G. Lüders, *"On the Equivalence of Invariance under Time Reversal and under Particle-Anti-particle Conjugation for Relativistic Field Theories"* [*Dansk. Mat. Fys. Medd.* 28, 5 (1954)], which referred to an unpublished remark by B. Zumino. The final contribution to the CPT Theorem was given by R. Jost, in *"Eine Bemerkung zum CPT Theorem"* [*Helv. Phys. Acta* 30, 409 (1957)], who showed that a weaker condition, called "weak local commutativity" was sufficient for the validity of the CPT Theorem.

[49] *Experimental Observation of Antideuteron Production*
T. Massam, Th. Muller, B. Righini, M. Schneegans, and A. Zichichi, *Nuovo Cimento* 39, 10 (1965).

[50] *The Discovery of Nuclear Antimatter and the Origin of the AMS Experiment*
S.S.C. Ting, in Proceedings of the *Symposium to celebrate the 30th anniversary of the Discovery of Nuclear Antimatter*, L. Maiani and R.A. Ricci (eds), Conference Proceedings 53, 21, Italian Physical Society, Bologna, Italy (1995).

[51] After the experimental discovery of the effect [1], W.F. Weisskopf and his student J. Bruce French were the first to correctly calculate the energy difference between the two energy levels of the Hydrogen Atom $(2^2S_{1/2} - 2^2P_{1/2})$. Julian Schwinger and Richard Feynman were also engaged in the same calculations but both of them had made the same mistake, thus getting the same (wrong) answer. Unfortunately (for Weisskopf), both Schwinger and Feynman were in contact with Weisskopf who could not believe that these two extremely bright members of the younger generation of physicists engaged in computing this unexpected new "effect" could both be wrong. Thus Weisskopf decided to postpone the publication of his (correct) result. Meanwhile Lamb and his student Norman Kroll published their (correct) result [*"On the Self-Energy of a Bound Electron"*, N.M. Kroll and W.E. Lamb, *Phys. Rev.* 75, 388 (1949)] while Weisskopf was waiting for the cross-check [*"The Electromagnetic Shift of Energy Levels"*, J.B. French and V.F. Weisskopf, *Phys. Rev.* 75, 1240 (1949)].

[52] The first report on "scaling" was presented by J.I. Friedman at the 14th International Conference on *High Energy Physics* in Vienna, 28 August-5 September 1968. The report was presented as paper n. 563 but not published in the Conference Proceedings. It was published as a SLAC preprint. The SLAC data on scaling were included in the Panofsky general report to the Conference where he says «... the apparent success of the parametrization of the cross-sections in the variable v/q^2 in addition to the large cross-section itself is at least indicative that point-like interactions are becoming involved». *"Low q^2 Electrodynamics, Elastic and Inelastic Electron (and Muon) Scattering"* W.K.H. Panofsky in Proceedings of 14th International Conference on *High Energy Physics* at Vienna 1968, J. Prentki and J. Steinberger (eds), page 23, published by CERN (1968). The following physicists participated in the inelastic electron scattering experiments: W.B. Atwood, E. Bloom, A. Bodek, M. Breidenbach,

G. Buschhorn, R. Cottrell, D. Coward, H. DeStaebler, R. Ditzler, J. Drees, J. Elias, G. Hartmann, C. Jordan, M. Mestayer, G. Miller, L. Mo, H. Piel, J. Poucher, C. Prescott, M. Riordan, L. Rochester, D. Sherden, M. Sogard, S. Stein, D. Trines, and R. Verdier. For additional acknowledgements see J.I. Friedman, H.W. Kendall and R.E. Taylor, *"Deep Inelastic Scattering: Acknowledgements"*, *Les Prix Nobel 1990*, (Almqvist and Wiksell, Stockholm/Uppsala 1991), also *Rev. Mod. Phys.* <u>63</u>, 629 (1991). For a detailed reconstruction of the events see J.I. Friedman *"Deep Inelastic Scattering Evidence for the Reality of Quarks"* in *"History of Original Ideas and Basic Discoveries in Particle Physics"*, H.B. Newman and T. Ypsilantis (eds), Plenum Press, New York and London, 725 (1994).

[53] *Quark Search at the ISR*
T. Massam and A. Zichichi, *CERN preprint*, June 1968;

Search for Fractionally Charged Particles Produced in Proton-Proton Collisions at the Highest ISR Energy
M. Basile, G. Cara Romeo, L. Cifarelli, P. Giusti, T. Massam, F. Palmonari, G. Valenti and A. Zichichi, *Nuovo Cimento* <u>40A</u>, 41 (1977); and

A Search for quarks in the CERN SPS Neutrino Beam
M. Basile, G. Cara Romeo, L. Cifarelli, A. Contin, G. D'Alì, P. Giusti, T. Massam, F. Palmonari, G. Sartorelli, G. Valenti and A. Zichichi, *Nuovo Cimento* <u>45A</u>, 281 (1978).

[54] *New Developments in Elementary Particle Physics*
A. Zichichi, *Rivista del Nuovo Cimento* <u>2</u>, n. 14, 1 (1979). The statement on page 2 of this paper, «*Unification of all forces needs first a Supersymmetry. This can be broken later, thus generating the sequence of the various forces of nature as we observe them*», was based on a work by A. Petermann and A. Zichichi in which the renormalization group running of the couplings using supersymmetry was studied with the result that the convergence of the three couplings improved. This work was not published, but perhaps known to a few. The statement quoted is the first instance in which it was pointed out that supersymmetry might play an important role in the convergence of the gauge couplings. In fact, the convergence of three straight lines (α_1^{-1} α_2^{-1} α_3^{-1}) with a change in slope is guaranteed by the Euclidean geometry, as long as the point where the slope changes is tuned appropriately. What is incorrect about the convergence of the couplings is that, with the initial conditions given by the LEP results, the change in slope needs to be at $M_{SUSY} \sim 1$ TeV as claimed by some authors not aware in 1991 of what was known in 1979 to A. Petermann and A. Zichichi.

[55] *The Effective Experimental Constraints on M_{SUSY} and M_{GUT}*
F. Anselmo, L. Cifarelli, A. Petermann and A. Zichichi, *Nuovo Cimento* <u>104A</u>, 1817 (1991).

[56] *The Simultaneous Evolution of Masses and Couplings: Consequences on Supersymmetry Spectra and Thresholds*
F. Anselmo, L. Cifarelli, A. Petermann and A. Zichichi, *Nuovo Cimento* <u>105A</u>, 1179 (1992).

[57] *Comparison of Grand Unified Theories with Electroweak and Strong Coupling Constants Measured at LEP*
U. Amaldi, W. de Boer and H. Fürstenau, *Phys. Lett.* B260, 447 (1991).

[58] *Interaction of Elementary Particles. Part I.*
H. Yukawa, *Proc. Physico-Math. Soc. Japan* 17, 48 (1935); and
Models and Methods in the Meson Theory
H. Yukawa, *Rev. Mod. Phys.* 21, 474 (1949).

[59] *Cloud Chamber Observations of Cosmic Rays at 4300 Meters Elevation and Near Sea Level*
C.D. Anderson and S.H. Neddermeyer, *Phys. Rev.* 50, 263 (1936).

[60] *On the Disintegration of Negative Mesons*
M. Conversi, E. Pancini and O. Piccioni, *Phys. Rev.* 71, 209 (1947).

[61] *Sui Mesoni dei Raggi Cosmici*
G. Puppi, *Nuovo Cimento* 5, 587 (1948). In this paper G. Puppi suggests that all Fermi processes could be described by the same coupling. In fact the decay rates of three different processes (π decay), (μ capture) and (μ decay) were found to be "approximately" the same. This is the origin of the Puppi triangle

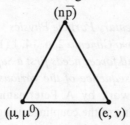

where the three vertices allow, through their couplings, to describe all known weak processes at that time. Note that Puppi distinguishes the neutral counter part of the muon, μ_0, (now known as ν_μ) from the neutral counter part of the electron (now called ν_e). The existence of a second neutrino, i.e. $\nu_\mu \neq \nu_e$, was established in 1962 by L.M. Lederman, M. Schwartz, J. Steinberger and Collaborators at BNL [62]. Other physicists have contributed to the universality of the Fermi interactions [63, 64, 65].

[62] *Observations of High-Energy Neutrino Reactions and the Existence of Two Kinds of Neutrinos*
G. Danby, J.-M. Gaillard, K. Goulianos, L.M. Lederman, N. Mistry, M. Schwartz and J. Steinberger, *Phys. Rev. Lett.* 9, 36 (1962).

[63] *Mesons and Nucleons*
O. Klein, *Nature* 161, 897 (1948).

[64] *Interaction of Mesons with Nucleons and Light Particles*
T.D. Lee, M. Rosenbluth and C.N. Yang, *Phys. Rev.* 75, 905 (1949).

[65] *Energy Spectrum of Electrons from Meson Decay*
J. Tiomno and J.A. Wheeler, *Rev. Mod. Phys.* 21, 144 (1949).

[66] *The Anomalous Magnetic Moment of the Muon*
G. Charpak, F. Farley, R.L. Garwin, T. Muller, J.C. Sens, V.L. Telegdi, C.M. York and A. Zichichi, Proceedings of the International Conference on *High-Energy Physics*, Rochester, NY, USA, 25 August-1 September 1960, Univ. Rochester, 778 (1960).

[67] *Measurement of the Anomalous Magnetic Moment of the Muon*
G. Charpak, F.J. Farley, R.L. Garwin, T. Muller, J.C. Sens, V.L. Telegdi and A. Zichichi, *Phys. Rev. Lett.* 6, 128 (1961).

[68] *A New Measurement of the Anomalous Magnetic Moment of the Muon*
G. Charpak, F.J. Farley, R.L. Garwin, T. Muller, J.C. Sens and A. Zichichi, *Phys. Lett.* 1, 16 (1962).

[69] *(g−2) and Its Consequences*
G. Charpak, F.J. Farley, R.L. Garwin, T. Muller, J.C. Sens and A. Zichichi
Proceedings of the International Conference on *High-Energy Physics*, Geneva, Switzerland, 4-11 July 1962, 476 (CERN, Geneva, 1962).

[70] *The Anomalous Magnetic Moment of the Muon*
G. Charpak, F.J. Farley, R.L. Garwin, Th. Muller, J.C. Sens and A. Zichichi, *Nuovo Cimento* 37, 1241 (1965).

[71] *A Measurement of the μ^+ Lifetime*
F.J. Farley, T. Massam, T. Muller and A. Zichichi
Proceedings of the International Conference on *High-Energy Physics*, Geneva, Switzerland, 4-11 July 1962, 415 (CERN, Geneva, 1962); and

CERN Work on Weak Interactions
A. Zichichi, in the February 1964 Meeting of the Royal Society. *Proc. Roy. Soc.* A285, 175 (1965).

[72] *A Measurement of the e^+ Polarization in Muon Decay: the e^+ Annihilation Method*
A. Buhler, N. Cabibbo, M. Fidecaro, T. Massam, Th. Muller, M. Schneegans and A. Zichichi, *Phys. Lett.* 7, 368 (1963).

[73] *Charge-Dependence of Nuclear Forces*
N. Kemmer, *Proc. Cambridge Phil. Soc.* 34, 354 (1938); and

Quantum Theory of Einstein-Bose Particles and Nuclear Interaction
N. Kemmer, *Proc. Roy. Soc.* A166, 127 (1938).

[74] *The Multiple Production of Mesons*
H.W. Lewis, J.R. Oppenheimer and S.A. Wouthuysen, *Phys. Rev.* 73, 127 (1948).

[75] *The Neutral Mesons*
A.G. Carlson, J.E. Hooper and D.T. King, *Phil. Mag.* 41, 701 (1950).

[76] *High Energy Photons from Proton-Nucleon Collisions*
R. Bjorklund, W.E. Crandall, B.J. Moyer and H.F. York, *Phys. Rev.* 77, 213 (1950).

[77] *The Gamma-Ray Spectrum from the Absorption of π^--Mesons in Hydrogen*
W.K.H. Panofsky, R.L. Aamodt and H.F. York, *Phys. Rev.* 78, 825 (1950);
The Gamma-Ray Spectrum Resulting from Capture of Negative π-Mesons in Hydrogen and Deuterium
W.K.H. Panofsky, R.L. Aamodt and J. Hadley, *Phys. Rev.* 81, 565 (1951).

[78] *Recension: "The Origin of the Concept of Nuclear Forces" by L.M. Brown and H. Rechenberg, Institute of Physics Publishing, Bristol and Philadelphia, 1966.*
G. Ekspong, *NIM* A394, 273 (1997).

[79] *On Gauge Invariance and Vacuum Polarization*
J. Schwinger, *Phys. Rev.* 82, 664 (1951).

[80] *The Axial Vector Current in Beta Decay*
M. Gell-Mann and M. Lévy, *Nuovo Cimento* 16, 705 (1960).

[81] *On the Pseudovector Current and Lepton Decays of Baryons and Mesons*
Chou Kuang-Chao, *Soviet Physics JETP* 12, 492 (1961).

[82] *Current Algebra and Some non-Strong Mesonic Decays*
D.G. Sutherland, *Nucl. Phys.* B2, 433 (1967).

[83] *Theoretical Aspects of High Energy Neutrino Interactions*
M. Veltman, *Proc. Roy. Soc.* A301, 107 (1967).

[84] *A PCAC Puzzle: $\pi^0 \rightarrow \gamma\gamma$ in the σ-Model*
J.S. Bell and R. Jackiw, *Nuovo Cimento* A60, 47 (1969).

[85] *Axial-Vector Vertex in Spinor Electrodynamics*
S.L. Adler, *Phys. Rev.* 177, 2426 (1969).

[86] *Absence of Higher-Order Corrections in the Anomalous Axial-Vector Divergence Equation*
S.L. Adler and W.A. Bardeen, *Phys. Rev.* 182, 1517 (1969).

[87] *Anomalous Ward Identities in Spinor Field Theories*
W.A. Bardeen, *Phys. Rev.* 184, 1848 (1969).

[88] *Magnetic Monopoles in Unified Gauge Theories*
G. 't Hooft, *Nuclear Phys.* B79, 276 (1974).

[89] *Particle Spectrum in Quantum Field Theory*
A.M. Polyakov, *ZhETF Pis. Red.* 20, 430 (1974).

[90] *Pseudoparticle Solutions of the Yang-Mills Equations*
A.A. Belavin, A.M. Polyakov, A.S. Schwartz and Yu.S. Tyupkin, *Phys. Lett.* 59B, 85 (1975).

[91] *Computation of the Quantum Effects due to a four-Dimensional Pseudoparticle*
G. 't Hooft, *Phys. Rev.* D14, 3432 (1976); and *err. Phys. Rev.* D18, 2199 (1978).

[92] *Symmetry Breaking through Bell-Jackiw Anomalies*
G. 't Hooft, *Phys. Rev. Lett.* 37, 8 (1976).

[93] *How Instantons Solve the U(1) Problem*
G. 't Hooft, *Phys. Rept.* 142, 357 (1986).

[94] *On the Analysis of τ-Meson data and the Nature of the τ-Meson*
R.H. Dalitz, *Phil. Mag.* 44, 1068 (1953);

Isotopic Spin Changes in τ and θ Decay
R.H. Dalitz, *Proceedings of the Physical Society* A69, 527 (1956);

Present Status of τ Spin-Parity
R.H. Dalitz, Proceedings of the *Sixth Annual Rochester Conference on High Energy Nuclear Physics*, Interscience Publishers, Inc., New York, 19 (1956); for a detailed record of the events which led to the (θ–τ) puzzle see R.H. Dalitz "*Kaon Decays to Pions: the τ–θ Problem*" in "*History of Original Ideas and Basic Discoveries in Particle Physics*", H.B. Newman and T. Ypsilantis (eds), Plenum Press, New York and London, 163 (1994).

[95] *Behavior of Neutral Particles under Charge Conjugation*
M. Gell-Mann and A. Pais, *Phys. Rev.* 97, 1387 (1955).

[96] *Observation of Long-Lived Neutral V Particles*
K. Lande, E.T. Booth, J. Impeduglia, L.M. Lederman, and W. Chinowski, *Phys. Rev.* 103, 1901 (1956).

[97] *Anomalous Regeneration of K_1^0 Mesons from K_2^0 Mesons*
R. Adair, W. Chinowsky, R. Crittenden, L.B. Leipuner, B. Musgrave and F.T. Shively *Phys. Rev.* 132, 2285 (1963).

[98] C.S. Wu, T.D. Lee, N. Cabibbo, V.F. Weisskopf, S.C.C. Ting, C. Villi, M. Conversi, A. Petermann, B.H. Wiik and G. Wolf
The Origin of the Third Family, O. Barnabei, L. Maiani, R.A. Ricci and F. Roversi Monaco (eds), Rome (1997); and World Scientific (1998).

[99] M. Gell-Mann, "*The Eightfold Way — A Theory of Strong-Interaction Symmetry*", California Institute Technology Synchrotron Lab. Report 20 (1961); "*Derivation of Strong Interactions from a Gauge Invariance*", Y. Ne'eman, *Nuclear Phys.* 26, 222 (1961); The experimental confirmation, in 1964, by Samios and collaborators [100] of the existence of the missing member of the baryonic decuplet, Ω⁻, appeared to be, at the time, a triumph for the "eightfold way". The choice of the letter Ω, the last in the Greek alphabet, was due to the conviction that this particle was going to be the last

ever to be discovered. See also: M. Gell-Mann and Y. Ne'eman, *"The Eightfold Way"*, W.A. Benjamin inc., New York and Amsterdam (1964).

[100] *Observation of a Hyperon with Strangeness Minus Three*
V.E. Barnes, P.L. Connolly, D.J. Crennell, B.B. Culwick, W.C. Delaney, W.B. Fowler, P.E. Hagerty, E.L. Hart, N. Horwitz, P.V.C. Hough, J.E. Jensen, J.K. Kopp, K.W. Lai, J. Leitner, J.L. Lloyd, G.W. London, T.W. Morris, Y. Oren, R.B. Palmer, A.G. Prodell, D. Radojictc, D.C. Rahm, C.R. Richardson, N.P. Samios, J.R. Sanford, R.P. Shutt, J.R. Smith, D.L. Stonehill, R.C. Strand, A.M. Thorndike, M.S. Webster, W.J. Willis and S.S. Yamamoto, *Phys. Rev. Lett.* 12, 204 (1964).

[101] *A Schematic Model of Baryons and Mesons*
M. Gell-Mann, *Phys. Lett.* 8, 214 (1964);

Fractionally Charged Particles and SU_6
G. Zweig, CERN Report TH 401, (1964), and Erice Lecture 1964, in *"Symmetries in Elementary Particle Physics"*, A. Zichichi (ed), Academic Press, New York, London (1965).

[102] *Particle Physics for Nuclear Physicists*
H.J. Lipkin, in *Physique Nucléaire*, Les-Houches 1968, C. DeWitt and V. Gillet (eds), Gordon and Breach, N.Y., 585 (1969);

Triality, Exotics and the Dynamical Basis of the Quark Model
H.J. Lipkin, *Phys. Lett.* 45B, 267 (1973);

A Systematics of Hadrons in Subnuclear Physics
Y. Nambu, in *"Preludes in Theoretical Physics"*, A. de-Shalit, H. Feshbach and L. Van Hove (eds), North Holland Pub. Comp., Amsterdam, 133 (1966).

[103] *Spin and Unitary-Spin Independence in a Paraquark Model of Baryons and Mesons*
O.W. Greenberg, *Phys. Rev. Lett.* 13, 598 (1964).

[104] *Three-Triplet Model with Double SU(3) Symmetry*
M.Y. Han and Y. Nambu, *Phys. Rev.* 139B, 1006 (1965).

[105] *Advantages of the Color Octet Gluon Picture*
H. Fritzsch, M. Gell-Mann and H. Leutwyler, *Phys. Lett.* 47B, 365 (1973).

[106] *Can We Make Sense Out of "Quantum Chromodynamics"?*
G. 't Hooft, in *"The Whys of Subnuclear Physics"*, Erice 1977, A. Zichichi (ed), Plenum Press, New York and London, 943 (1978).

[107] *Gauge Theories with Unified, Weak, Electromagnetic and Strong Interactions*
G. 't Hooft, in *EPS Int. Conf. on High Energy Physics*, Palermo, 23-28 June 1975, A. Zichichi (ed), Editrice Compositori, Bologna, 1225 (1976).

[108] *Are Matter and Antimatter Symmetric?*
T.D. Lee, in Proceedings of the *Symposium to celebrate the 30th anniversary of the Discovery of Nuclear Antimatter*, L. Maiani and R.A. Ricci (eds), Conference Proceedings 53, page 1, Italian Physical Society, Bologna, Italy (1995).

[109] We discuss in this note how the mass difference, between a quark and its antiquark state, compares with the mass differences measured for particle antiparticle states (mesons and baryons). To study the mass differences between quark and antiquark states it is necessary to study the particle states made of three quarks (baryons), not those made of quark antiquark pairs (mesons).

Let us consider the best known mass differences:

$$| \Delta m_{K\bar{K}} | \lesssim 4 \times 10^{-10} \text{ eV/c}^2 ; \tag{1}$$

$$| \Delta m_{p\bar{p}} | \leq 40 \text{ eV/c}^2 . \tag{2}$$

They differ by eleven orders of magnitude. Nevertheless, the high precision result on $\Delta m_{K\bar{K}}$ can be of no value to establish if CPT holds in the quark antiquark masses. This point needs to be discussed in some detail.

The masses of the K^0 meson and of its antiparticle state $\bar{K}^0$ are, respectively:

$$m_{K^0} = m_d + m_{\bar{s}} + m_{d\bar{s}}^{Bag} + m_{d\bar{s}}^{Rad}$$

$$m_{\bar{K}^0} = m_{\bar{d}} + m_s + m_{\bar{d}s}^{Bag} + m_{\bar{d}s}^{Rad} \tag{1.1}$$

where

m_d	$\equiv$	the mass of the fundamental fermion "d".
m_s	$\equiv$	the mass of the fundamental fermion "s".
$m_{\bar{d}}$	$\equiv$	the mass of the fundamental antifermion "$\bar{d}$".
$m_{\bar{s}}$	$\equiv$	the mass of the fundamental antifermion "$\bar{s}$".
$m_{d\bar{s}}^{Bag}$	$\equiv$	the QCD-Bag energy needed to confine the pair "$d\bar{s}$".
$m_{\bar{d}s}^{Bag}$	$\equiv$	the QCD-Bag energy needed to confine the pair "$\bar{d}s$".
m^{Rad}	$\equiv$	radiative corrections.

Let us suppose that a source of CPT breaking exists at some energy scale. This is certainly true at the string unification scale (E_{SU}) where CPT invariance loses its foundations [T.D. Lee, Ref. 108].

If this happens, we might expect

$$m_{q_i} \neq m_{\bar{q}_i}$$

where m_{q_i} $\equiv$ the mass of the quark with flavour "i"

$m_{\bar{q}_i}$ $\equiv$ the mass of the antiquark with flavour "i".

Let us assume the simplest CPT breaking effect: i.e. all quark flavours differ from their antiquark states by the same amount Δm:

$$m_{q_i} = m_{\bar{q}_i} \pm \Delta m_i$$

(with i = 1, 2 ... 6) and no effects of CPT breaking are present in the Bag and in the Radiative parts. There are two possibilities: either $m_{q_i} > m_{\bar{q}_i}$ or $m_{q_i} < m_{\bar{q}_i}$. In both cases the final result would be the same.

Let us choose the first one:

$$m_{q_i} = m_{\bar{q}_i} - \Delta m_i$$

From the relations (1.1) we have

$$\Delta m_{K^0\bar{K}^0} = m_{K^0} - m_{\bar{K}^0} = (m_d - m_{\bar{d}}) + (m_{\bar{s}} - m_s) = \Delta m_d - \Delta m_s.$$

The simplest CPT breaking possibility is that all Δm_i are equal. In this case

$$\Delta m_{K^0 \bar{K}^0} = \text{zero.}$$

despite the fact that $m_{q_i} \neq m_{\bar{q}_i}$.

The very small limit on the mass difference between the K^0 meson and its antistate $\bar{K}^0$,

$$\Delta m_{K\bar{K}} \gtrsim 4 \times 10^{-10} \text{ eV/c}^2 \tag{1}$$

might have no effect on the value of the difference between quark and antiquark states. For this difference to be investigated, we need to compare the mass of a particle (made of three quarks) and of its antiparticle (made of three antiquarks). Let us consider the proton; its mass is given by

$$m_p = 2 m_u + m_d + m_{uud}^{Bag} + m_{uud}^{Rad}$$

where the symbols are self evident (and explained in the text). For the antiproton we have

$$m_{\bar{p}} = 2 m_{\bar{u}} + m_{\bar{d}} + m_{\bar{u}\bar{u}\bar{d}}^{Bag} + m_{\bar{u}\bar{u}\bar{d}}^{Rad}$$

Assuming, as before, the simplest CPT breaking mechanism, which only affects the fundamental fermion masses,

$$m_{q_i} \neq m_{\bar{q}_i} \quad \text{and} \quad m_{q_i} = m_{\bar{q}_i} - \Delta m_i$$

without effects on m^{Bag} and m^{Rad}, we have from the experimental results [111]

$$\Delta m_{p\bar{p}} = (m_p - m_{\bar{p}}) = (22 \pm 40) \text{ eV} = 3 \Delta m (q\bar{q}), \tag{2}$$

i.e.

$$\Delta m (q\bar{q}) \lesssim 20 \text{ eV/c}^2.$$

To sum up: there are two particle-antiparticle limits (1) and (2): $\Delta m_{K^0 \bar{K}^0}$ and $\Delta m_{p\bar{p}}$. The first one is 11 orders of magnitude better than the second one. These eleven orders of magnitude might be of no help in trying to establish if CPT holds for those mechanisms which provide the masses to the fundamental fermions called quarks. In other words, if a CPT violating effect on the quark-antiquark mass difference as large as

$$\Delta m_{q\bar{q}} = 20 \text{ eV/c}^2,$$

[which is more than ten orders of magnitude above the $\Delta m_{K\bar{K}}$ limit (1)] was the correct CPT breaking effect on the quark-antiquark mass difference, the result (1) could still hold true. In fact we have shown that the mass difference between mesons and antimesons might be zero even if the CPT breaking effects were very large on the mass difference between quarks and antiquarks.

[110] *The Antideuteron Experiment - Recollections of the fine times and closing remarks*
A. Zichichi, in Proceedings of the *Symposium to celebrate the 30th anniversary of the Discovery of Nuclear Antimatter*, L. Maiani and R.A. Ricci (eds), Conference Proceedings <u>53</u>, page 123, Italian Physical Society, Bologna, Italy (1995).

[111] *The Discovery of Nuclear Antimatter*
L. Maiani and R.A. Ricci (eds), Conference Proceedings 53, Italian Physical Society, Bologna, Italy (1995).

[112] e^+e^- *Annihilation into Hadrons at SPEAR*
G.J. Feldman, in *E.P.S. Int. Conf. on High Energy Physics*, Palermo, 23-28 June 1975, A. Zichichi (ed), Editrice Compositori, Bologna, 233 (1976).

[113] *The Basic Steps which Led to the Discovery of the Heavy Lepton τ: a Historical Record*
C. Villi, in "*The Origin of the Third Family*", O. Barnabei, L. Maiani, R.A. Ricci and F. Roversi Monaco (eds), Rome (1997); and World Scientific (1998).

[114] *Experimental Observation of a Heavy Particle J*
J.J. Aubert, U. Becker, P.J. Biggs, J. Burger, M. Chen, G. Everhart, P. Goldhagen, J. Leong, T. McCorriston, T.G. Rhoades, M. Rohde, S.C.C. Ting, S.L. Wu and Y.Y. Lee, *Phys. Rev. Lett.* 33, 1404 (1974).

[115] *Discovery of a Narrow Resonance in e^+e^- Annihilation*
J.-E. Augustin, A.M. Boyarski, M. Breidenbach, F. Bulos, J.T. Dakin, G.J. Feldman, G.E. Fischer, D. Fryberger, G. Hanson, B. Jean-Marie, R.R. Larsen, V. Lüth, H.L. Lynch, D. Lyon, C.C. Morehouse, J.M. Paterson, M.L. Perl, B. Richter, P. Rapidis, R.F. Schwitters, W.M. Tanenbaum, F. Vannucci, G.S. Abrams, D. Briggs, W. Chinowsky, C.E. Friedberg, G. Goldhaber, R.J. Hollebeck, J.A. Kadyk, B. Lulu, F. Pierre, G.H. Trilling, J.S. Whitaker, J. Wiss and J.E. Zipse, *Phys. Rev. Lett.* 33, 1406 (1974).

[116] *First Search for Sequential Heavy Leptons at ADONE*
A. Zichichi, CERN-PPE/93-58 and CERN/LAA/93-18, 2 April 1993. Presented at the Symposium on "The τ particle", in honour of Martin Perl's 65th birthday, SLAC, Stanford, CA, USA, 24 July 1992; and in the Proceedings of the Summer Institute on Particle Physics "*The Third Family and the Physics of Flavor*", L. Vassilllian (ed), SLAC CONF-9207140 UC-414 (T/E), 603 (1993).

[117] *Observation of a Dimuon Resonance at 9.5 GeV in 400-GeV Proton-Nucleus Collisions*
S.W. Herb, D.C. Hom, L.M. Lederman, J.C. Sens, H.D. Snyder, J.K. Yoh, J.A. Appel, B.C. Brown, C.N. Brown, W.R. Innes, K. Ueno, T. Yamanouchi, A.S. Ito, H. Jöstlein, D.M. Kaplan and R.D. Kephart, *Phys. Rev. Lett.* 39, 252 (1977);

Observation of Structure in the Y Region
W.R. Innes, J.A. Appel, B.C. Brown, C.N. Brown, K. Ueno, T. Yamanouchi, S.W. Herb, D.C. Hom, L.M. Lederman, J.C. Sens, H.D. Snyder, J.K. Yoh, R.J. Fisk, A.S. Ito, H. Jöstlein, D.M. Kaplan, and R.D. Kephart, *Phys. Rev. Lett.* 39, 1240 (1977);

A Volume in honour of A. Zichichi
on the Occasion of the Galvani Bicentenary Celebrations

Alma Mater Studiorum
Saecularia Nona

ACCADEMIA DELLE SCIENZE

THE CREATION OF
QUANTUM CHROMODYNAMICS
and
The EFFECTIVE ENERGY

V.N. Gribov, G. 't Hooft,
G. Veneziano and V.F. Weisskopf

Edited by
L.N. Lipatov

Published by
the University of Bologna and its Academy of Sciences — Bologna 1998 — Galvani Celebrations

Vladimir N. Gribov

«Each pair of interacting particles, when producing systems consisting of many hadronic particles, had its own final state. No-one knew how to settle this flagrant contradiction.

When I read the paper "Evidence of the same multiparticle production mechanism in (pp) collisions as in (e⁺e⁻) annihilation", I realized that something very interesting had been found. In fact the introduction of the "Effective Energy" in the analysis of (pp) collisions at the CERN-ISR gave a totally unexpected result.»

Gerardus 't Hooft

«Theoreticians were unable to prescribe what experimentalists had to look for to establish the universal nature of these final interactions. The experimental results were discouraging; scattering experiments yielded different final states for each pair of interacting particles. So it happened that these experimental aspects of QCD had to wait until experimentalists themselves came with the right idea. The showers come with what is now called an "effective energy", and, in terms of this quantity, universality could be established.»

Gabriele Veneziano

«The introduction of the "effective energy" concept allowed Nino to show that pairs of interacting particles, whether the interaction is strong, electromagnetic or weak, produce multiparticle hadronic states with the same basic properties. This provides an essential experimental support to the fundamental QCD process that is believed to lie at the basis of hadron production: the hadronization of quarks and gluons.»

Victor F. Weisskopf

«Nino had the right idea at the right time. In fact, the Universality Features are experimentally hidden insofar as the "Effective Energy" is not introduced in the analysis.

It is for me very gratifying that the discovery of the "Effective Energy" was implemented at the CERN-ISR. Moreover, I like it because it still has to be quantitatively explained by QCD.»

Reproduction of Reference 123: front and cover pages.

"Evidence for the Y" and a Search for New Narrow Resonances
K. Ueno, B.C. Brown, C.N. Brown, W.R. Innes, R.D. Kephart, T. Yamanouchi, S.W. Herb, D.C. Hom, L.M. Lederman, H.D. Snyder, J.K. Yoh, R.J. Fisk, A.S. Ito, H. Jöstlein and D.M. Kaplan, *Phys. Rev. Lett.* 42, 486 (1979).

[118] A record of the annual evolution in the multitude of baryonic and mesonic states can be found in the proceedings of the Erice Schools [16]. An example of the proliferation in the meson resonances is *"Meson Resonances and Related Electromagnetic Phenomena"*, R.H. Dalitz and A. Zichichi (eds), Editrice Compositori, Proceedings of the EPS Conference, Bologna (1971).

[119] *The End of a Myth: High-P_T Physics*
M. Basile, J. Berbiers, G. Cara Romeo, L. Cifarelli, A. Contin, G. D'Alì, C. Del Papa, P. Giusti, T. Massam, R. Nania, F. Palmonari, G. Sartorelli, M. Spinetti, G. Susinno, L. Votano and A. Zichichi, in *"Quarks, Leptons, and their Constituents"*, Erice 1984, A. Zichichi (ed), Plenum Press, New York and London, 1 (1988).

[120] *"Leading" Physics at LHC Including Machine Studies Plus Detector R&D (LAA)*
T. Taylor, H. Wenninger and A. Zichichi, *Nuovo Cimento* 108A, 1477 (1995);
and in *"Vacuum and Vacua - The Physics of Nothing"*, Erice 1995, A. Zichichi (ed), World Scientific, 381 (1996).

[121] *The "Leading"-Baryon Effect in Strong, Weak, and Electromagnetic Interactions*
M. Basile, G. Cara Romeo, L. Cifarelli, A. Contin, G. D'Alì, P. Di Cesare, B. Esposito, P. Giusti, T. Massam, R. Nania, F. Palmonari, V. Rossi, G. Sartorelli, M. Spinetti, G. Susinno, G. Valenti, L. Votano and A. Zichichi, *Lettere al Nuovo Cimento* 32, 321 (1981).

[122] *What Can We Learn From High-Energy, Soft (pp) Interactions*
M. Basile, G. Bonvicini, G. Cara Romeo, L. Cifarelli, A. Contin, M. Curatolo, G. D'Alì, B. Esposito, P. Giusti, T. Massam, R. Nania, F. Palmonari, A. Petrosino, V. Rossi, G. Sartorelli, M. Spinetti, G. Susinno, G. Valenti, L. Votano and A. Zichichi in *"The Unity of the Fundamental Interactions"*, Erice 1981, A. Zichichi (ed), Plenum Press, New York and London, 695 (1983).

[123] For a complete set of references concerning this topic see *"The Creation of Quantum ChromoDynamics and the Effective Energy"* V.N. Gribov, G. 't Hooft, G. Veneziano and V.F. Weisskopf; N.L. Lipatov (ed), Academy of Sciences and University of Bologna, INFN, SIF, published by World Scientific, 1998.

[124] *The Gran Sasso Laboratory and the Eloisatron Project*
A. Zichichi, in *"Old and New Forces of Nature"*, Erice 1985, A. Zichichi (ed), Plenum Press, New York and London, 335 (1988).

[125] *Evidence for η' Leading*
L. Cifarelli, T. Massam, D. Migani and A. Zichichi, in *"Highlights: 50 Years Later"*, Erice 1997, A. Zichichi (ed), World Scientific (1998). See also D. Migani *Thesis*, July 1997, Bologna University.

[126] *Field Theories with "Superconductor" Solutions*
J. Goldstone, *Nuovo Cimento* <u>19</u>, 154 (1961).

[127] *Observation of a Nonstrange Meson of Mass 959 MeV*
G.R. Kalbfleisch, L.W. Alvarez, A. Barbaro-Galtieri, O.I. Dahl, P. Eberhard, W.E. Humphrey, J.S. Lindsey, D.W. Merrill, J.J. Murray, A. Rittenberg, R.R. Ross, J.B. Shafer, F.T. Shively, D.M. Siegel, G.A. Smith and R.D. Tripp, *Phys. Rev. Lett.* <u>12</u>, 527 (1964);

Existence of a New Meson of Mass 960 MeV
M. Goldberg, M. Gundzik, S. Lichtman, J. Leitner, M. Primer, P.L. Connolly, E.L. Hart, K.W. Lai, G. London, N.P. Samios and S.S. Yamamoto, *Phys. Rev. Lett.* <u>12</u>, 546 (1964).

[128] *Evidence for a New Decay Mode of the X^0-Meson: $X^0 \rightarrow 2\gamma$*
D. Bollini, A. Buhler-Broglin, P. Dalpiaz, T. Massam, F. Navach, F.L. Navarria, M.A. Schneegans and A. Zichichi, *Nuovo Cimento* <u>58A</u>, 289 (1968).

[129] *The Decay Mode $\omega \rightarrow e^+e^-$ and a Direct Determination of the $\omega - \phi$ Mixing Angle*
D. Bollini, A. Buhler-Broglin, P. Dalpiaz, T. Massam, F. Navach, F.L. Navarria, M.A. Schneegans and A. Zichichi, *Nuovo Cimento* <u>57A</u>, 404 (1968); see also

Observation of the Rare Decay Mode of the ϕ-Meson: $\phi \rightarrow e^+e^-$
D. Bollini, A. Buhler-Broglin, P. Dalpiaz, T. Massam, F. Navach, F.L. Navarria, M.A. Schneegans and A. Zichichi, *Nuovo Cimento* <u>56A</u>, 1173 (1968).

[130] *Radiative Decays of Mesons*
A joint publication: University of Bologna and Academy of Sciences. To be published.

The Basic SU(3) Mixing: $\omega_8 \longleftrightarrow \omega_1$
A. Zichichi, in *"Evolution of Particle Physics"*, Academic Press Inc., New York and London, 299 (1970).

[131] *Measurement of the Branching Ratio $\Gamma(X^0 \rightarrow \gamma\gamma) / \Gamma(X^0 \rightarrow TOTAL)$*
P. Dalpiaz, P.L. Frabetti, T. Massam, F.L. Navarria and A. Zichichi, *Phys. Lett.* <u>42B</u>, 377 (1972).

[132] *The Convergence of the Gauge Couplings at E_{GUT} and Above: Consequences for $\alpha_3(M_Z)$ and SUSY Breaking*
F. Anselmo, L. Cifarelli, A. Petermann and A. Zichichi, *Nuovo Cimento* <u>105A</u>, 1025 (1992).

[133] *Analytic Study of the Supersymmetry-Breaking Scale at Two Loops*
F. Anselmo, L. Cifarelli, A. Petermann and A. Zichichi, *Nuovo Cimento* <u>105A</u>, 1201 (1992); and

The full Two-Loop Approach to the Problem of the Light Supersymmetric Threshold
A. Petermann and A. Zichichi, *Nuovo Cimento* <u>108A</u>, 105 (1995).

[134] *Understanding Where the Supersymmetry Threshold Should Be*
A. Zichichi, in Proceedings of the Workshop on *"Ten Years of SUSY Confronting*

Experiment", CERN, Geneva, Switzerland, 7-9 September 1992, CERN-TH 6707/92-PPE/92-180, 94. CERN-PPE/92-149 and CERN/LAA/MSL/92-017, 7 September 1992.

[135] *New Precision Electroweak Tests of SU(5) × U(1) Supergravity*
J.L. Lopez, D.V. Nanopoulos, G.T. Park and A. Zichichi, *Phys. Rev.* D49, 4835 (1994).

[136] *The Simplest, String-Derivable, Supergravity Model and its Experimental Predictions*
J.L. Lopez, D.V. Nanopoulos and A. Zichichi, *Phys. Rev.* D49, 343 (1994).

[137] *Acoplanar Di-Leptons and Mixed Events on the Basis of two Supergravity Model Predictions*
F. Anselmo, G. Anzivino, F. Arzarello, G. Bari, M. Basile, T. Barillari, L. Bellagamba, J. Berbiers, R. Bertin, F. Block, D. Boscherini, G. Bruni, P. Bruni, G. Cara Romeo, M. Chiarini, L. Cifarelli, F. Cindolo, F. Ciralli, A. Contin, I. Crotty, C. D'Ambrosio, M. Dardo, S. De Pasquale, R. De Salvo, L. Fava, F. Frasconi, P. Ford, P. Giusti, T. Gys, D. Hatzifotiadou, M. Hourican, M. Kaur, G. Iacobucci, G. La Commare, J. Lamas Valverde, H. Larsen, G. Laurenti, H. Leutz, G. Levi, J.L. Lopez, G. Maccarrone, A. Margotti, M. Marino, T. Massam, A. Menshikov, C. Musso, R. Nania, D.V. Nanopoulos, C. Nemoz, D. Panzieri, A. Peterman, D. Piedigrossi, D. Puertolas, H. Pois, S. Qian, E. Ruffino, G. Sartorelli, I. Schipper, J. Seguinot, S. Tailhardat, R. Timellini, M. Vivargent, X. Wang, M.C.S. Williams, T. Ypsilantis and A. Zichichi, *Nuovo Cimento* 106A, 1389 (1993).

[138] *Experimental Aspects of SU(5) × U(1) Supergravity*
J.L. Lopez, D.V. Nanopoulos, G.T. Park, X. Wang and A. Zichichi, *Phys. Rev.* D50, 2164 (1994).

[139] *A String no-Scale Supergravity Model and its Experimental Consequences*
J.L. Lopez, D.V. Nanopoulos and A. Zichichi, *Phys. Rev.* D52, 4178 (1995).

[140] *No-Scale Supergravity Confronts LEP Diphoton Events*
J.L. Lopez, D.V. Nanopoulos and A. Zichichi, October 1996 - *hep-ph/9610235*.

[141] *A Study of the Various Approaches to M_{GUT} and α_{GUT}*
F. Anselmo, L. Cifarelli and A. Zichichi, *Nuovo Cimento* 105A, 1335 (1992).

[142] *A χ^2-Test to Study the α_1, α_2, α_3 Convergence for High-Precision LEP data, Having in Mind the SUSY Threshold*
F. Anselmo, L. Cifarelli and A. Zichichi, *Nuovo Cimento* 105A, 1357 (1992).

[143] *Speech by the Mayor of Erice on the Occasion of the Dedication of the Richard P. Feynman Lecture Hall*
in *History of Original Ideas and Basic Discoveries in Particle Physics*, H.B. Newman and T. Ypsilantis (eds), Erice 1994, Plenum Press, New York and London, Vol. B352, xxi (1996).

156

SPEECH BY THE MAYOR OF ERICE ON THE OCCASION OF THE DEDICATION OF THE RICHARD P. FEYNMAN LECTURE HALL

Ladies and Gentlemen, Dear Professor Zichichi,

I am indeed honoured and have great pleasure in welcoming you all in the name of the town of Erice. You represent here the highest scientific level of Modern Physics through your inventions and your discoveries; these have determined the greatest scientific conquests during the past fifty years.

I mentioned my pleasure not just as a formality, but because of an extraordinary fact which has occurred: a discovery of major artistic significance in the Hall which will shortly be dedicated to the great physicist Richard Feynman who was among the very first teachers of this Institution: in fact he came here thirty years ago, in 1964. The Centre which I have the honour of presiding, is very grateful to Professor Feynman and to all those illustrious scientists who with their precious collaboration have contributed to the success and the prestige of this Centre. Well over thirty years have gone by since Professor Antonino Zichichi set up this Centre for Scientific Culture dedicated to Ettore Majorana. In these past thirty-three years sixty-thousand nine-hundred and sixty-eight Scientists, coming from nine-hundred and twenty-five Universities of one-hundred and thirty-nine Countries have taken part in the Activities of the Erice Schools. For all this I would like to thank Professor Zichichi. But there is yet another reason, once more he has given us a marvellous gift: the discovery of a Last Supper at San Rocco. A masterpiece which was lying beneath many layers of whitewash and which is dated around 1500.

A detail which you may not know about. The Zichichi family came to Erice in 1200, seven hundred years ago. The first Zichichi in Erice was a magistrate. In 1500 one of Professor Zichichi's ancestors, Brother Ludovico Zichichi, founded the Poor Man's Morsel, an Institution created to help the poor who were arriving in Erice looking for help.

Another one of his ancestors, also named Antonino, gave a substantial part of his fortune to the San Rocco Monastery. It is this particular donation which has given life and lustre to San Rocco.

With these memories in mind, the town of Erice today renews to Professor Zichichi and to his family who have been here in Erice for seven centuries, the profound feelings of gratitude and affection. To all of you, in particular to the Scientific Director Professor Harvey Newman and to the Director of the Workshop Professor Thomas Ypsilantis, most hearty thanks for having once again shown a token of consideration and of friendship to this ancient town.

xxi

Reproduction of page xxi [143].

[144] *On a Relativistically Invariant Formulation of the Quantum Theory of Wave Fields*
S. Tomonaga, *Prog. Theoret. Phys.* 1, 27 (1946);

On a Relativistically Invariant Formulation of the Quantum Theory of Wave Fields. II.
Z. Koba, T. Tati, and S. Tomonaga, *Prog. Theoret. Phys.* 2, 101, 198 (1947);

On a Relativistically Invariant Formulation of the Quantum Theory of Wave Fields. V.
S. Kanesawa and S. Tomonaga, *Prog. Theoret. Phys.* 3, 101 (1948);

On Infinite Field Reactions in Quantum Field Theory
S. Tomonaga, *Phys. Rev.* 74, 224 (1948).

[145] *On Quantum-Electrodynamics and the Magnetic Moment of the Electron*
J. Schwinger, *Phys. Rev.* 73, 416 (1948);

Quantum Electrodymanics. I. A Covariant Formulation
J. Schwinger, *Phys. Rev.* 74, 1439 (1948).

[146] *Space-Time Approach to Non-Relativistic Quantum Mechanics*
R.P. Feynman, *Rev. Mod. Phys.* 20, 367 (1948);

A Relativistic Cut-Off for Classical Electrodynamics
R.P. Feynman, *Phys. Rev.* 74, 939 (1948);

Relativistic Cut-Off for Quantum Electrodynamics
R.P. Feynman, *Phys. Rev.* 74, 1430 (1948);

Interaction with the Absorber as the Mechanism of Radiation
J.A. Wheeler and R.P. Feynman, *Rev. Mod. Phys.* 17, 157 (1945).

[147] *The Radiation Theories of Tomonaga, Schwinger, and Feynman*
F.J. Dyson, *Phys. Rev.* 75, 486 (1949).

[148] *The S Matrix in Quantum Electrodynamics*
F.J. Dyson, *Phys. Rev.* 75, 1736 (1949).

[149] Proceedings of the 12th Solvay Conference on *"The Quantum Theory of Fields"*,
University of Brussels, October 1961. Interscience Publishers, New York (1961).

[150] *The Normalization Group in Quantum Theory*
E.C.G. Stueckelberg and A. Petermann, *Helv. Phys. Acta* 24, 317 (1951);

La Normalisation des Constantes dans la Théorie des Quanta
E.C.G. Stueckelberg and A. Petermann, *Helv. Phys. Acta* 26, 499 (1953);

Introduction to the Theory of Quantized Fields
N.N. Bogoliubov and D.V. Shirkov, Interscience Publishers, New York (1959);

Renormalization Group and the Deep Structure of the Proton
A. Petermann, *Phys. Reports* 53, 157 (1979).

[151] *Quantum Electrodynamics at Small Distances*
M. Gell-Mann and F.E. Low, *Phys. Rev.* 95, 1300 (1954).

[152] For a lucid description of the subject see *"Renormalization and Symmetry: a review for
Non-Specialists"*, S. Coleman, in *"Properties of the Fundamental Interactions"*, Erice

1971, A. Zichichi (ed), Editrice Compositori, Bologna, 605 (1973); see also
Renormalization
J.C. Collins, Cambridge University Press, (1984).

[153] *Regularization and Renormalization of Gauge Fields*
G. 't Hooft and M. Veltman, *Nuclear Phys.* B44, 189 (1972).

[154] *Lowest Order "Divergent" Graphs in v-Dimensional Space*
C.G. Bollini and J.J. Giambiagi, *Phys. Lett.* 40B, 566 (1972);

A Method of Gauge-Invariant Regularization
J.F. Ashmore, *Lettere al Nuovo Cimento* 4, 289 (1972).

[155] *Secret Symmetry. An Introduction to Spontaneous Symmetry Breakdown and Gauge Fields*
S. Coleman, in *"Laws of Hadronic Matter"*, Erice 1973, A. Zichichi (ed), Academic Press, New York and London, 139 (1975).

[156] *Renormalization of Massless Yang-Mills Fields*
G. 't Hooft, *Nuclear Phys.* B33, 173 (1971).

[157] *Renormalizable Lagrangians for Massive Yang-Mills Fields*
G. 't Hooft, *Nuclear Phys.* B35, 167 (1971).

[158] *Combinatorics of Gauge Fields*
G. 't Hooft and M. Veltman, *Nucl. Phys.* B50, 318 (1972).

[159] *The Superworld I*
Erice 1986, A. Zichichi (ed), Plenum Press, New York and London, (1990);

The Superworld II
Erice 1987, A. Zichichi (ed), Plenum Press, New York and London, (1990);

The Superworld III
Erice 1988, A. Zichichi (ed), Plenum Press, New York and London, (1990).

[160] *All You need to Know about the Higgs Boson*
L. Maiani, Proceedings Ecole d'Ete de Physique des Particules, Gif-Sur-Yvette, 45 (1979).

[161] *Under the Spell of the Gauge Principle*
G. 't Hooft, World Scientific (1994).

[162] *Fundamental Problems*
L.D. Landau, in M. Fierz and V.F. Weisskopf (eds), *Theoretical Physics in the Twentieth Century; a Memorial Volume to Wolfgang Pauli.* Interscience Publishers, New York, 245-248 (1960);

On Point Interaction in Quantum Electrodynamics
L.D. Landau and I. Pomeranchuk, *Dokl Akad. Nauk SSSR* 102, 489 (1955);

On the Quantum Theory of Fields, L.D. Landau in W. Pauli (ed), *Niels Bohr and the Development of Physics*, Pergamon Press, New York, 52-69 (1955). For a discussion of the basic problems see A.S. Wightman *"Should we Believe in Quantum Field Theory?"* in *The Whys of Subnuclear Physics*, Erice 1977, A. Zichichi (ed), Plenum Press, New York and London, 983 (1979).

[163] *Partial-Symmetries of Weak Interactions*
S.L. Glashow, *Nuclear Phys.* 22, 579 (1961).

[164] *Quantum Theory of Gravitation*
R.P. Feynman, *Acta Phys. Polonica* 24, 697 (1963).

[165] *Unitarity and Causality in a Renormalizable Field Theory with Unstable Particles*
M. Veltman, *Physica* 29, 186 (1963);

Perturbation Theory of Massive Yang-Mills Fields
M. Veltman, *Nuclear Phys.* B7, 637 (1968);

Massive Yang-Mills Fields
J. Reiff and M. Veltman, *Nuclear Phys.* B13, 545 (1969);

Generalized Ward Identities and Yang-Mills Fields
M. Veltman, *Nuclear Phys.* B21, 288 (1970);

Massive and Mass-Less Yang-Mills and Gravitational Fields
H. van Dam and M. Veltman, *Nuclear Phys.* B22, 397 (1970).

[166] *Electromagnetic and Weak Interactions*
A. Salam and J.C. Ward, *Phys. Lett.* 13, 168 (1964);

A Model of Leptons
S. Weinberg, *Phys. Rev. Lett.* 19, 1264 (1967);

Weak and Electromagnetic Interactions
A. Salam, Nobel Symposium 1968, N. Svartholm (ed), Almqvist and Wiksell, Wiley Interscience, 367 (1968).

[167] *Renormalization of the Abelian Higgs-Kibble Model*
C. Becchi, A. Rouet and R. Stora, *Commun. Math. Phys.* 42, 127 (1975);

Renormalization of Gauge Theories
C. Becchi, A. Rouet and R. Stora, *Annals of Physics* 98, 287 (1976);

I.V. Tyutin, *Lebedev Prepr. FIAN* 39 (1975), unpublished.

[168] *Unitary Symmetry and Leptonic Decays*
N. Cabibbo, *Phys. Rev. Lett.* 10, 531 (1963).

[169] *Leptonic Decays and the Unitary Symmetry*
N. Cabibbo, in *"Strong Electromagnetic and Weak Interactions"*, Erice 1963, A. Zichichi (ed), W.A. Benjamin Inc, New York and Amsterdam, 191 (1964).

[170] *Consequences of SU3 Symmetry in Weak Interactions*
R.P. Feynman, in *"Symmetries in Elementary Particle Physics"*, Erice 1964, A. Zichichi (ed), Academic Press, New York and London, 111 (1965).

[171] M. Gell-Mann and M. Levy, *Nuovo Cimento* 16, 705 (1960); in a foot-note of this paper (already quoted [80] and where the authors present the σ–model) the authors suggest that a parameter can be associated with the "strange" currents in order not to spoil the universality of the Fermi coupling.

[172] *Weak Interactions with Lepton-Hadron Symmetry*
S.L. Glashow, J. Iliopoulos and L. Maiani, *Phys. Rev.* D2, 1285 (1970).

[173] *CP-Violation in the Renormalizable Theory of Weak Interaction*
M. Kobayashi and T. Maskawa, *Prog. Theor. Phys.* 49, 652 (1973).

[174] *Kamiokande and Super-Kamiokande*
M. Koshiba, in *"From the Planck Length to the Hubble Radius"*, Erice 1998, World Scientific, to be published.

[175] *Speakable and Unspeakable in Quantum Mechanics*
J.S. Bell, Cambridge University Press, London (1987).

[176] *Current Algebra and Anomalies*
S.B. Treiman, R. Jackiw, B. Zumino and E. Witten (eds), pages 81 and 211, World Scientific.

[177] *Vacuum Periodicity in a Yang-Mills Quantum Theory*
R. Jackiw and C. Rebbi, *Phys. Rev. Lett.* 37, 172 (1976); for a clear Lecture on the topic see *"The uses of Instantons"* S. Coleman, in *"The Whys of Subnuclear Physics"*, Erice 1978, A. Zichichi (ed), Plenum Press, New York and London, 805 (1978).

[178] *The Structure of the Gauge Theory Vacuum*
C.G. Callan, R.F. Dashen and D.J. Gross, *Phys. Lett.* 63B, 334 (1976).

[179] *The uses of Instantons*
S. Coleman, in *"The Whys of Subnuclear Physics"*, Erice 1977, A. Zichichi (ed), Plenum Press, New York and London, 805 (1979).

[180] *The INFN Eloisatron Project*
A. Zichichi, in Proceedings of the HARC '93 International Worskop, World Scientific, 363 (1994).

[181] *Eloisatron (The European LOng Intersecting Storage Accelerator)*
M. Basile, J. Berbiers, G. Bonvicini, E. Boschi, N. Cabibbo, G. Cara Romeo, L. Cifarelli, M. Civita, A. Contin, M. Curatolo, G. D'Alì, M. Dardo, C. Del Papa, B. Esposito, L. Ferrario, M.I. Ferrero, S. Galassini, P. Giusti, I. Laakso, A.R. Leo, M. Leo, G. Luches, P. Lunardi, A. Marino, T. Massam, R. Nania, V. Nassisi, F. Palmonari, M. Puglisi, F. Resmini, C. Rizzuto, P. Rotelli, G. Sartorelli, G. Soliani, M. Spinetti, L. Stringa, G. Susinno, S. Tazzari, L. Votano and A. Zichichi
INFN/AE-83/7, June 1983; INFN/AE-84/2, January 1984 (revised version 1985); Presented at the Galileo Galilei and Alfred B. Nobel Celebrations "Science for Peace", Sanremo and Rome, Italy, 1-11 May 1983.

[182] *The INFN Eloisatron Project*
C. Aglietta, C. Alberini, G. Badino, G. Bari, M. Basile, M. Bassetti, J. Berbiers, A. Bertin, E. Boschi, V. Braginski, R. Bruzzese, N. Cabibbo, G. Cara Romeo, R. Casaccia, C. Castagnoli, A. Castellina, A. Castelvetri, L. Cifarelli, F. Cindolo, M. Civita, A. Contin, G. D'Alì, M. Dardo, C. Del Papa, V. De Sabbata, L. Ferrario, W. Fulgione, S. Galassini, P. Galeotti, M. Gasperini, P. Giusti, R. Goldoni, G. Iacobucci, I. Laakso, A.R. Leo, M. Leo, G. Luches, G. Maccarrone, A. Marino, T. Massam, V.N. Melnikov, R. Meunier, F. Motta, R. Nania, V. Nassisi, G. Navarra, F. Palmonari, G. Papini, G. Passotti, P. Pelfer, G. Pocci, G. Prisco, M. Puglisi, M. Ricci, G. Rinaldi, C. Rizzuto, F. Rohrbach, P. Rotelli, O. Saavedra, N. Sacchetti, G. Sartorelli, G. Soliani, M. Spadoni, M. Steuer, G. Susinno, S. Tazzari, K. Thorne, G. Torelli, G.C. Trinchero, P. Vallania, G. Venturi, S. Vernetto, F. Villa, A. Vitale, L. Votano, M. Willutzky and A. Zichichi, in Proceedings of the INFN Eloisatron Workshop, Plenum Press, New York, 297 (1988); and RECFA, CERN, Geneva, Switzerland, 19 June 1986.

[183] *The Eloisatron Project: Eurasian LOng Intersecting Storage Accelerator*
A. Zichichi, in Proceedings on *"New Aspects of High-Energy Proton-Proton Collisions"*, Plenum Press, New York and London, 1 (1989).

[184] *Physics up to 200 TeV*
Erice 1990, A. Zichichi (ed), Plenum Press, New York and London, (1991).

[185] *Eloisatron: New Strategies for Supercolliders*
A. Zichichi, in Proceedings of the 9th Workshop of the INFN Eloisatron Project *"Perspectives for New Detectors in Future Supercolliders"*, World Scientific, 238 (1991).

[186] *Why 200 TeV*
A. Zichichi, in Proceedings of the 12th Workshop of the INFN Eloisatron Project, Plenum Press, New York and London, 1 (1991).

[187] *The INFN Eloisatron Project*
A. Zichichi, CERN-PPE/93-62 and CERN/LAA/93-20, 13 April 1993.

[188] *The Eloisatron Project:*
A. Zichichi, in *"The Superworld II"*, Erice 1987, A. Zichichi (ed), Plenum Press, New York and London, 443 (1990).

[189] *The Lepton Asymmetry Analyser*
C. Alberini, G. Bari, M. Basile, J. Berbiers, G. Cara Romeo, R. Casaccia, L. Cifarelli, F. Cindolo, A. Contin, G. D'Alì, C. Del Papa, S. De Pasquale, G. Iacobucci, I. Laakso, T.D. Lee, G. Maccarrone, T. Massam, R. Meunier, F. Motta, R. Nania, F. Palmonari, E. Perotto, G. Prisco, F. Rohrbach, P. Rotelli, G. Sartorelli, G. Susinno, L. Votano, M. Willutzky and A. Zichichi, *CERN/SPSC* 86-3, SPSC/P200 Add. 1, March 1986; *INFN/AE-86/4*, March 1986; *CERN/SPSC* 86-18, SPSC/P200 Add. 2, May 1986.

[190] *The LAA Project, Report n. 1*
A. Zichichi, et al., *CERN/LAA*, 15 December 1986.

[191] *The LAA Project, Report n. 2*
A. Zichichi, et al., *CERN/LAA*, 25 June 1987.

[192] *The LAA Project*
G. Anzivino, G. Bari, M. Basile, U. Becker, J. Berbiers, R.K. Boch, G. Cara Romeo, R. Casaccia, G. Charpak, L. Cifarelli, F. Cindolo, G. Comby, A. Contin, G. D'Alì, C. Del Papa, S. De Pasquale, B. Guerard, H. Heijne, R. Horisberger, G. Iacobucci, G. Jarlskog, P. Jarron, W.M. Kelly, J. Kirkby, I. Laakso, T.D. Lee, H. Leutz, G. Maccarrone, J. Malos, T. Massam, R. Meunier, P. Mine, G. Mork, F. Motta, R. Nania, F. Palmonari, E. Perotto, G. Prisco, F. Rohrbach, P. Rotelli, G. Sartorelli, F. Sauli, D.H. Saxon, D. Scigocki, P. Schlein, M. Suffert, G. Susinno, M. Vivargent, L. Votano, W. Wallraff, R. Wigmans, M. Willutzky, K. Winter, F. Wittgenstein and A. Zichichi, *CERN-EP*/87-122, 14 July 1987.

[193] *The LAA Project*
G. Anzivino, G. Bari, M. Basile, U. Becker, J. Berbiers, R.K. Boch, G. Cara Romeo, R. Casaccia, G. Charpak, L. Cifarelli, F. Cindolo, G. Comby, A. Contin, G. D'Alì, C. Del Papa, S. De Pasquale, B. Guerard, H. Heijne, R. Horisberger, G. Iacobucci, G. Jarlskog, P. Jarron, W.M. Kelly, J. Kirkby, I. Laakso, T.D. Lee, H. Leutz, G. Maccarrone, J. Malos, T. Massam, R. Meunier, P. Mine, G. Mork, F. Motta, R. Nania, F. Palmonari, E. Perotto, G. Prisco, F. Rohrbach, P. Rotelli, G. Sartorelli, F. Sauli, D.H. Saxon, D. Scigocki, P. Schlein, M. Suffert, G. Susinno, M. Vivargent, L. Votano, W. Wallraff, R. Wigmans, M. Willutzky, K. Winter, F. Wittgenstein and A. Zichichi, *ICFA* - Instrumentation Bulletin 3, September 1987.

[194] *The LAA Project, Report n. 3*
A. Zichichi, et al., *CERN/LAA*, 19 November 1987.

[195] *Perspectives for a New Detector at a Future Supercollider: the LAA Project*
A. Zichichi, et al., in *"Heavy Flavours and High-Energy Collisions in the 1-100 TeV Range"*, Plenum Press, New York and London, 357 (1989).

[196] *The LAA Project, Report n. 4*
A. Zichichi, et al., *CERN/LAA*/88-1, 25 July 1988.

[197] *The LAA Project*
G. Anzivino et al., *Rivista del Nuovo Cimento* 13, n. 5 (1990).

[198] *The LAA Project: Second Year of Activity*
A. Zichichi, in *"The Challenging Questions"*, Erice 1989, A. Zichichi (ed), Plenum Press, New York and London, 221 (1990).

[199] *Advances in Technology for High-Energy Subnuclear Physics: Contribution of the LAA project*
D. Acosta, et al., *Rivista del Nuovo Cimento* 13, n. 10-11 (1990).

[200] *The Main Achievements of the LAA Project, Report N°7*
A. Zichichi, et al., *CERN/LAA/91-1*, 1 March 1991.

[201] *The Main Achievements of the LAA Project*
A. Zichichi, in *"Physics up to 200 TeV"*, Erice 1990, A. Zichichi (ed), Plenum Press, New York and London, 327 (1991).

[202] *The Monte Carlo Simulation Laboratory (MSL) of LAA*
F. Anselmo, F. Block, G. Brugnola, L. Cifarelli, E. Eskut, D. Hatzifotiadou, G. La Commare, C. Maidantchik, M. Marino, S. Qian, Yu.M. Shabelski, G. Xexeo, Y.Ye and A. Zichichi, *CERN/DRDC* 92-44, LAA Status Report, 3 September 1992.

[203] *Predictions for Secondary Particle Production at Existing and Future Hadron-Hadron Colliders*
F. Anselmo, L. Cifarelli, E. Eskut and Yu.M. Shabelski, *Nuovo Cimento* 105A, 1371 (1992).

[204] *Charm and Beauty Hadroproduction Models: QGSM vs. Lund*
L. Cifarelli, E. Eskut and Yu.M. Shabelski, *Nuovo Cimento* 106A, 389 (1993).

[205] *Neural Networks for Higgs Search*
F. Anselmo, F. Block, G. Brugnola, L. Cifarelli, D. Hatzifotiadou, G. La Commare and M. Marino, *Nuovo Cimento* 107A, 129 (1994).

[206] *Heavy Higgs Search with Hadron Supercolliders up to $\sqrt{s} = 200$ TeV*
F. Anselmo et al., *Nuovo Cimento* 107A, 783 (1994).

[207] *The End of Superworld III*
S.L. Glashow, in Proceedings *"The Super World III"*, Erice 1988, A. Zichichi (ed), Plenum, New York and London, 411 (1990);

Particle Physics in the Nineties
S.L. Glashow, in *"Physics up to 200 TeV"*, Erice 1990, A. Zichichi (ed), Plenum, New York and London, 1 (1991).

[208] *The Glorious Future of Particle Physics*
D.J. Gross, in Proceedings *"From Superstring to Present-day Physics"*, Erice 1994, A. Zichichi (ed), World Scientific, 1 (1995).

[209] *The Physical Vacuum as a Condensate*
T.D. Lee, in Proceedings *Effective Theories and Fundamental Interactions*, Erice 1996, A. Zichichi (ed), World Scientific, 3 (1997).

[210] *Beyond the Standard Model*
F. Wilczek, Presented at the 35th Course of the "Ettore Majorana" International School of Subnuclear Physics, Erice, Italy, 26 August-4 September 1997.

[211] *The Limits of our Imagination in Elementary Particle Theory*
G. 't Hooft, Presented at the 35th Course of the "Ettore Majorana" International School of Subnuclear Physics, Erice, Italy, 26 August-4 September 1997.

[212] As early as 1938 E.C.G. Stueckelberg introduced what is now called the "Baryon number conservation". Stueckelberg noted that the number of protons and neutrons (the Heavy particles) in the Universe can never change, otherwise matter itself would be unstable. This postulate became of great relevance with the advent of the Grand Unified Theories in the seventies.

[213] *Scienza ed Emergenze Planetarie - Il Paradosso dell'Era Moderna*
A. Zichichi, Rizzoli (1st Edition 1993, 3rd Edition 1994), Supersaggi Bur Rizzoli (1st Edition 1996, 6th Edition 1998).

[214] *L'Infinito*
A. Zichichi, Rizzoli-Bur (1st Edition 1988, 7th Edition 1994) and a more recent edition by Pratiche Editrice (1998).

[215] *My own testimony on the validity of 't Hooft's statement in experimental physics*
It is not enough to have an original idea. Lord Patrick Maynard Stuart Blackett says "We experimentalists are not like theorists: the originality of an idea is not for being printed in a paper, but for being shown in the implementation of an original experiment." This is the first sentence I have chosen from the set selected by the Director of the Dirac Museum. Dirac is an example of a theorist who does not limit himself to presenting a new idea with a few pages of computation (which often confuse the reader). Dirac brought his logical reasoning to the extreme consequences. Thus, both in theory and in experiment, the progress of physics is due to those who have the perseverance of not only having an original idea, but of investigating its logical structure in terms of its consequences. Here follow a few examples from my own experience.
The third lepton, ([17] and § II.2-2, page 50). In the late fifties, I realised that if the pion mass was not what it was, the muon had very little chance of being so obviously present everywhere; and if a new lepton of 1 GeV mass (or heavier) would have been there, no-one would have seen it; I did not limit myself to discussing this topic with a few colleagues; I followed Blackett's teaching. And this is how I realised that the best "signature" for a heavy lepton would have been "eμ" acoplanar pairs; this is how I invented the "preshower" to improve "electron" identification by an order of magnitude; this is why I studied how to improve "muon" identification; this is how I experimentally established that the best production mechanism could not be ($p\bar{p}$), but (e^+e^-) annihilation.
Another example. Matter-Antimatter Symmetry, ([49, 50] and § II.2-1, page 41). In the sixties, the need to check the symmetry between matter and antimatter came to the limelight. In fact the apparent triumph of the S-matrix theory to describe strong interactions and the violation of the "well-established" symmetry operators (C, P, CP, T) in weak interactions and in the K-meson decay physics, together with the discovery of scaling in Deep Inelastic Scattering (DIS) and the non-breaking of the protons in high energy collisions, appeared to put in serious difficulty the basic structure of all relativistic quantum field theories, and therefore the validity of the celebrated CPT theorem (see § II.1-1, page 23; § II.2-2, pages 43, 49; and page 128 in Ref. 50). On the other hand, the basic reason why nuclear antimatter had to exist was CPT.
But the first example of nuclear antimatter, the antideuteron, had been searched for and found not to be there. I did not limit myself to saying that it would have been

important to build a beam of negatively charged "partially separated" particles in order to have a very high intensity. I did not limit myself to suggesting a very advanced electronic device in order to increase, by an order of magnitude, the accuracy for time-of-flight (TOF) measurements. I did bring all my ideas to the point of full implementation in a detailed experiment, where the antideuteron was found, thus proving nuclear matter-antimatter symmetry, and so credence could be given to CPT and to RQFT.

Another problem of concern in the physics of strong interactions was the "mixing" problem in meson physics, ([128, 129, 130, 131] and page 82). It was necessary to know why the vector mesons (ρ, ω, ϕ) did not show the same behaviour as the pseudoscalar mesons (π, η, η'). I did not limit myself to saying that the best way to study this problem was to measure with the best possible accuracy the electromagnetic decay rates of the vector mesons (the (e^+e^-) colliders did not yet exist), ($\rho \to e^+e^-$), ($\omega \to e^+e^-$) ($\phi \to e^+e^-$), and to see if the heaviest meson (known at that time with the symbol X^0) was decaying into two γ's ($X^0 \to \gamma\gamma$). These were times when experimental physics was dominated by bubble chambers. I designed and built a non-bubble-chamber detector, NBC [130]; it consisted of an original neutron missing mass spectrometer (page 82) coupled with a powerful electromagnetic detector which allowed to clearly identify all final states of the decaying mesons which consisted of (e^+e^-) pairs or $(\gamma\gamma)$ pairs. The mass of the meson (be it pseudoscalar or vector) was measured by the neutron missing mass spectrometer.

When in 1968 I heard Pif (W.K.H.) Panofsky [52] reporting in Vienna on (ep) deep-inelastic-scattering, the so-called Bjorken scaling, whose immediate consequence was that the "partons" inside a proton behaved as "free" particles, I did not limit myself to saying that it would have been interesting to check if, in violent (pp) collisions, "free" partons were produced. Since the "partons" were suspected to be the quarks suggested earlier by M. Gell-Mann and G. Zweig (we now know that partons can also be gluons), the experiment needed was the search for fractionally charged particles in the final states of the violent (pp) interactions at the ISR. To perform the experiment, a new type of plastic scintillator was needed; with very long attenuation length since the counters had to be put inside a very big magnet. These scintillators did not exist on the market. We studied the problem and built the most powerful and sensitive scintillators. The result was that no quarks were produced, despite the violent (pp) collisions, ([53] and page 29).

When the physics of strong interactions finally became the physics of quarks and gluons, QCD had the problem, defined by Gribov [123] as being its "hidden side": i.e., the enormous number of different final states produced by different pairs of interacting particles, such as (π^-p, pp, $\bar{p}p$, Kp, e^+e^-, νp, μp, ep, etc.). I did not limit myself to suggesting that a totally different approach was needed to put all these final states on the same basis. I found what this basis could be: and this is how the "Effective Energy" became the correct quantity to be measured in each interaction (see page 58). To begin this study, it was necessary to analyze tens of thousands of (pp) interactions at the ISR. And this was done despite all the difficulties to be overcome.

So, when a problem appears in its value as a key issue to be faced, the only way out is to bring the logical reasoning — be it of experimental, theoretical or technical nature — to the deepest level of consequences. This is how progress is made in advanced research.

[216] *The Validity of 't Hooft's Statement in Theoretical Physics*

The most spectacular example for the validity of 't Hooft's statement, in the domain of theoretical physics, is the work of Gerardus 't Hooft himself in the early seventies. More exactly in 1971, when he proved that *realistic* gauge theoretical models for electro-weak interactions could be renormalized. This was the result of deep theoretical analysis brought to the extreme consequences. Thanks to the understanding of the renormalization processes and rules, the few models left would yield predictions as precise as in the theory for the electromagnetic interactions between electrons and photons (quantum electrodynamics). This came as a complete surprise. In fact, during many decades all attempts to describe weak interactions with massive bosons had failed. The importance of this development was that, all of a sudden, the rules for building such models could be accurately formulated.

The rules required that:

1) there is a limited number of *elementary* particles, all with spin = 0, 1/2 or at most 1;

2) the *interactions* among these particles are subject to a restricted set of possibilities (renormalizable interactions); in technical jargon this means that all interactions among elementary particles must be of limited dimensionality (less or equal to four, in a 4-dimensional space-time).

Finally, and most importantly:

3) the masses of various particle species must be introduced via what is now called the "Higgs mechanism" (more precisely: Higgs-Englert-Brout).

It was quickly realized that there was a small number of alternative possibilities for the real world that obeyed these restrictive requirements. From the few available options, the choice could be made by doing new experiments. It soon turned out that the symmetry structure was SU(2) × U(1) already proposed in 1961 by S. Glashow [163] and the interactions were the ones suggested in 1967 by S. Weinberg [166], with a few additions needed to correctly incorporate the hadronic sector; and this accurately represented the real situation. It is thanks to the rigorous work of G. 't Hooft that a vast number of testable predictions could be made; most notably on the existence of neutral current interactions among leptons as well as baryons, and on a new species of quark, named "charm". Theory, for instance, put restrictions on the mass of the charmed quark, and these predictions were corroborated by experiment.

The techniques for calculating the higher-order effects, and thus to obtain the required accuracies in the predictions, were subsequently greatly improved by the so-called "dimensional renormalization" procedure ('t Hooft and Veltman, 1972, [153, 154], see also page 91). Later, the model allowed for more predictions such as the top and beauty quarks. Before these were actually observed, theoreticians had two reasons to suspect their existence. One was that, in renormalizable theories, these quarks would provide most naturally for a mechanism for the observed violation of CP symmetry; the second reason was that the newly discovered third lepton (1974) (now called τ) would have caused an anomaly that would have obstructed renormalizability unless it was cancelled out by a new generation of quarks, the top and the beauty (see page 35). The newly acquired ability to renormalize gauge theories was also essential for the construction of models for the strong interactions. It turned out to be necessary, first to realize that, after renormalization, quantized gauge field theory exhibits a property named "asymptotic freedom". This property, long thought to be impossible, was necessary to explain a phenomenon called "Bjorken scaling", first observed at SLAC in violent (ep) collisions. As we have seen in the text, the key-point here was the negative sign of the β function. This was first published by D. Gross and F. Wilczek in 1973, but it had been announced publicly — though not in writing — by 't Hooft a year earlier at a conference in Marseilles. It led to the gauge theory of the "coloured forces" (quantum chromodynamics).

Asymptotic freedom by itself, however, does not explain one of the most striking features of the strong force, which is the experimentally well-established fact (ISR-1968-74) that the constituent particles, the quarks, do not come out in (pp) interactions, despite the very high energy of the collision; the quarks appear to be held together by infinite potential wells (permanent quark confinement).

Several physicists contributed to understanding this feature, but the clearest picture of the situation was provided by 't Hooft when he showed that it can be attributed to Bose condensation of colour-magnetic charges. The existence of isolated magnetic charges in a gauge theory (magnetic monopoles) had been discovered by 't Hooft in 1974. Although particles with single magnetic charges reacting to *Maxwell's* magnetic fields have not (yet) been discovered, and probably will never show up, they do play important roles in theories of the early universe, as well as in numerous mathematical studies of gauge theories in general, and quantum chromodynamics in particular. *Colour*-magnetic charges are important structures that help theorists understand the results produced in computer simulations of quantum chromodynamics on a lattice, and these simulations in turn enable to understand the observed spectrum of hadronic particles. The exact spectrum is still not there, but progress is being made day after day in trying to understand the non-perturbative properties of QCD.

One of the properties is the experimentally established "Zweig's rule": mesons tend to interact mainly in those channels where quark quantum numbers are directly exchanged. Here again we find G. 't Hooft, who discovered that the theory of the strong force simplifies if the system of three different "colours" is replaced by one with a very large number, N, of colours. The $1/N$ expansion gives a set of "planar" Feynman diagrams. This sheds further light on the string nature of the forces confining quarks, and it implies that the mutual interactions among mesons are proportional to $1/N$. In addition, those interactions not obeying Zweig's rule are suppressed by more $1/N$ factors.

Early calculations in quantum chromodynamics still showed one important discrepancy with experimental observations: the properties of a pseudo-scalar isoscalar meson, now called the η, and its SU(3) partner, η'. These objects appeared to violate a symmetry called chiral U(1) symmetry. The primary equations of QCD, as laid down in its Lagrangian, appeared to imply the conservation of the vector current associated to chiral U(1) symmetry, but this should require the masses of η and η' to be much lower than actually measured, and it would also not allow the mixing between η, η' and π^0 to be as strong as it was experimentally observed. All these features were explained at once when 't Hooft found a class of new tunnelling phenomena, which he called "Instantons", that explicitly violate chiral U(1) conservation [see § II.3-5.1]. Instantons [see § II.3-5.2] now play a crucial role in the analysis of gauge theories and their super-symmetric extensions. They then showed up in condensed matter physics, for example, in recent theories concerning the fractional quantum Hall effect.

It was after these fundamental discoveries that the models for the electro-weak and the strong forces could be combined in what is now called "the Standard Model". The Standard Model implied a revolution in our thinking about elementary particles. No particle accelerator experiment, and no theoretical treatment of elementary particles, can presently be imagined without a prior comparison with, and analysis of, the Standard Model predictions. The Standard Model itself could not have been formulated at all if we had not known the rules for its renormalization.

Just because the present picture of elementary particles in the energy domain that can be reached by experiments is so clear, many theoreticians have switched to the new problems that can now be addressed: the extreme short distance aspects of elementary particles, and eventually the incorporation of the gravitational force. These activities have as yet not led to any predictions that can soon be tested, but they did provide many new insights. One of these is that, at the ultra-short distance scales, black holes must begin to play an important role. Another is the so-called "holographic principle", which is the fact that the dynamical degrees of freedom for fundamental particles, when the gravitational force is included, must somehow be distributed on the boundary of the system, rather than in a volume. The reason is that, if we would try to put more information in this part of the universe, that information would disappear in a black hole, because information costs energy and energy causes gravitational attraction. As a consequence there is a limit on the amount of information one can have in a given region. An implication of this theory is that the universe looks as if a picture was taken using the principle of "holography". The resulting picture would not be sharp. Since the data are distributed on a surface, instead of inside a volume, the resulting holographic picture of the universe is "blurred". These principles, as they were put forward by 't Hooft, are presently being vigorously investigated in the context of string theory.

Neither string theory, nor supergravity and other approaches to quantize the gravitational field would have been possible if theoreticians had not learned from 't Hooft's work how to deal with virtual states, ghost particles, unitarity and causality when he found the rules for renormalizing gauge theories.

[217] *A Few Words on the Standard Model*

We have seen how subnuclear physics was born in 1947, how it has developed during this half-a-century and where it will possibly go from its present status.

Its present status, the Standard Model, is the most formidable synthesis of all phenomena known to exist, from the inner structure of a proton to the extreme border of the universe. We have seen in Fig. II.3.1 (page 89) that the basic steps of the Standard Model are five. Let us add a few words of more detailed nature. The Standard Model is based on:

i)   matter particles (quarks or leptons), fermions, and therefore described by the Dirac spinor function ψ; these particles have spin $\frac{1}{2}$;

ii) gauge particles [the photon (γ), the weak bosons $(W^{\pm}, Z^0)$ and the gluons (g)], all being of vector nature; these particles have spin 1;

iii) scalar particles, with imaginary masses; these particles have zero spin. They cause Spontaneous Symmetry Breaking (SSB) [see § II.3-3], thus providing real masses to matter particles (spin $\frac{1}{2}$) and gauge particles (spin 1) in addition to themselfes (spin zero particles), called Higgs particles. We denote these particles by the letter H.

All fundamental interactions in the Standard Model are based on simple Feynman diagrams of which the one below is, at the three level, the prototype:

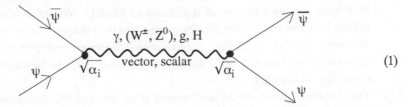

$$\gamma, (W^\pm, Z^0), g, H \qquad (1)$$

The basic ingredient of the Standard Model is the covariant derivative, which must have a term for each local gauge invariance; in fact, the way a gauge force is generated is through the substitution of the standard derivative ∂_μ with the covariant derivative D_μ.

Since the local gauge invariances are U(1), SU(2) and SU(3), there will be three gauge couplings g_1, g_2, g_3, respectively, with

$$\alpha_i = \frac{g_i^2}{4\pi} \ (i = 1, 2, 3).$$

Each symmetry group has its generators

$$
\begin{aligned}
U(1) &\rightarrow Y \\
SU(2) &\rightarrow \tau_i = \tau_1, \tau_2, \tau_3 \\
SU(3) &\rightarrow \lambda_a = \lambda_1, \lambda_2 .... \lambda_8 .
\end{aligned}
$$

Each gauge invariance generates a gauge boson

$$
\begin{aligned}
U(1) &\rightarrow B_\mu \\
SU(2) &\rightarrow W_\mu^i \\
SU(3) &\rightarrow G_\mu^a
\end{aligned}
$$

The covariant derivative of the Standard Model is therefore:

$$D_\mu = \partial_\mu - i\, g_1 \frac{Y}{2}\, B_\mu - i\, g_2 \frac{\tau^i}{2}\, W_\mu^i - i\, g_3 \frac{\lambda^a}{2}\, G_\mu^a \ .$$

As we have seen, the photon described by the field A_μ is the result of a mixing between the field B_μ and the neutral component of the weak boson field, W_μ^3; the values of the gauge couplings are not given by theory but by experiment, as well as all masses. More precisely, the Standard Model needs the following inputs:

3 Gauge couplings: g_1, g_2, g_3.

1 Non-perturbative vacuum-angle parameters: θ_3 for the non-Abelian gauge forces SU(3).

6 Quark masses (which range from few MeV up to 180 GeV for the top quark).

3 Charged lepton masses (which range from 0.5 MeV up to 1.77 GeV for the third lepton).

3 Neutral lepton masses (which are very small, not zero, and with very small mass differences $\Delta m^2 \simeq 10^{-3}\,\text{eV}^2$).

4 Mixing angle parameters in the quark sector (which come from the experimentally measured transitions across quark flavour states, including a phase angle to describe CP violation in the flavour-changing charged currents).

4   (Presumably) mixing angle parameters in the lepton sector.

2   Weak boson mass parameters (which can be taken as $M_{W^\pm}$ and M_{Higgs}).

These 26 parameters are the reason why we need to search for new physics beyond the Standard Model. But, how could this search be started if we were unable to find a firm basis for the structure of the Standard Model? We have seen that this could indeed be achieved and that the key-source was the effective theoretical understanding of renormalization. Note that, if gravitational forces are included, there are two more parameters in a theory beyond the Standard Model: i.e. the Newton constant and the cosmological constant.

The Standard Model would not be there if no one had had the courage to think in a logical way and to carry logical "thinking" to its extreme consequences.

Graham M. Shore, Ludvig D. Faddeev, Yakov Azimov, Masatoshi Koshiba, the author, Edward Witten, Bjorn H. Wiik, Sheldon L. Glashow and Gerardus 't Hooft at Erice (1998).

APPENDIX
List of the Erice Subnuclear Physics Series and of its Main Topics

From the Foreword.

«The reconstruction of the first five decades of subnuclear physics has a strong link with the "Ettore Majorana" School of Subnuclear Physics at Erice, a small town on the top of a mountain founded — according to the myth — by the son of Venus. Here, every year since 1963, the development of subnuclear physics has been recorded and the hottest topics of the moment were registered as faithfully as possible in the discussion sessions of the Erice School. At this School, I have attempted to have as Lecturers the most active and authoritative members of the subnuclear physics community. Their ingenuity, their wisdom, their rigorous attempts to understand the constituents and the fundamental forces of nature are reported in the volumes of the Subnuclear Physics Series. The original ideas which have flourished during the past 50 years were the focus of an intense intellectual activity, both for theorists and for experimentalists, who have contributed to the lectures and to the discussion sessions of the Erice Subnuclear Physics School.»

The volumes of the Subnuclear Series are listed according to their time sequence. This list is followed by the Subject Index[*] which illustrates the main topics. When a subject is discussed in several courses, the order in the list is chronological. The key is as follows: topic, author, page, volume number of the Erice School, year.

The Subject Index could be of interest for those readers who would like to deepen their knowledge of a topic mentioned in the present volume.

(*)   The subject index has been prepared by my student, Fabrizio Pierella.

John Stewart Bell in Erice (1975).

J.S. BELL

Speakable and
unspeakable
in quantum
mechanics

CAMBRIDGE UNIVERSITY PRESS

For the great Nino
At last a little
in return for so
much.

John
1987 Dec 7

THE SUBNUCLEAR SERIES

vol. 01.-Erice 1963- **'STRONG, ELECTROMAGNETIC, AND WEAK INTERACTIONS',**
W. A. Benjamin Inc., New York (1964)

vol. 02.-Erice 1964- **'SYMMETRIES IN ELEMENTARY PARTICLE PHYSICS',**
Academic Press, New York and London (1965)

vol. 03.-Erice 1965- **'RECENT DEVELOPMENTS IN PARTICLE SYMMETRIES',**
Academic Press, New York and London (1966)

vol. 04.-Erice 1966- **'STRONG AND WEAK INTERACTIONS',**
Academic Press, New York and London (1967)

vol. 05.-Erice 1967- **'HADRONS AND THEIR INTERACTIONS',**
Academic Press, New York and London (1968)

vol. 06.-Erice 1968- **'THEORY AND PHENOMENOLOGY IN PARTICLE PHYSICS',**
Academic Press, New York and London (1969)

vol. 07.-Erice 1969- **'SUBNUCLEAR PHENOMENA',**
Academic Press, New York and London (1970)

vol. 08.-Erice 1970- **'ELEMENTARY PROCESSES AT HIGH ENERGY',**
Academic Press, New York and London (1971)

vol. 09.-Erice 1971- **'PROPERTIES OF THE FUNDAMENTAL INTERACTIONS',**
Editrice Compositori, Bologna (1972)

vol.09a.-Erice 1971- **'MESON RESONANCES AND RELATED ELECTROMAGNETIC PHENOMENA',**
R.H. Dalitz and A. Zichichi (eds)
Editrice Compositori, Bologna (1971)

vol.10.-Erice 1972- **'HIGHLIGHTS IN PARTICLE PHYSICS',**
Editrice Compositori, Bologna (1972)

vol.11.-Erice 1973- **'LAWS OF HADRONIC MATTER',**
Academic Press, New York and London (1974)

vol.12.-Erice 1974- **'LEPTON AND HADRON STRUCTURE',**
Academic Press, New York and London (1975)

vol.13.-Erice 1975- **'NEW PHENOMENA IN SUBNUCLEAR PHYSICS',**
Plenum Press, New York and London (1976)

vol.14.-Erice 1976- **'UNDERSTANDING THE FUNDAMENTAL CONSTITUENTS OF MATTER',**
Plenum Press, New York and London (1977)

vol.15.-Erice 1977- **'THE WHYS OF SUBNUCLEAR PHYSICS',**
Plenum Press, New York and London (1978)

vol.16.-Erice 1978- **'THE NEW ASPECTS OF SUBNUCLEAR PHYSICS',**
Plenum Press, New York and London (1979)

vol.16a.-Erice 1978- **'HADRONIC MATTER AT THE EXTREME ENERGY DENSITY',**
N. Cabibbo and L. Sertorio (eds)
Plenum Press, New York and London (1980)

vol.17.-Erice 1979- **'POINTLIKE STRUCTURES INSIDE AND OUTSIDE HADRONS',**
Plenum Press, New York and London (1980)

vol.17a.-Erice 1979- **'PROBING HADRONS WITH LEPTONS',**
G. Preparata and J.J. Aubert (eds)
Plenum Press, New York and London (1980)

vol.18.-Erice 1980- **'THE HIGH-ENERGY LIMIT',**
Plenum Press, New York and London (1981)

Mrs Goldhaber and Rudolf Peierls in Erice (1980).

Hans Bethe in Erice (1981).

Simon Van der Meer in Erice (1984).

Ken Wilson in Erice (1975).

vol.08a.-Erice 1980-	**'UNIFICATION OF THE FUNDAMENTAL PARTICLE INTERACTIONS'**, *S. Ferrara, J. Ellis and P. van Nieuwenhuizen (eds)* *Plenum Press, New York and London (1980)*
vol.18b.-Erice 1980-	**'NEUTRINO PHYSICS AND ASTROPHYSICS'**, *E. Fiorini (ed)* *Plenum Press, New York and London (1982)*
vol.19.-Erice 1981-	**'THE UNITY OF THE FUNDAMENTAL INTERACTIONS'**, *Plenum Press, New York and London (1982)*
vol.19a.-Erice 1981-	**'UNIFICATION OF THE FUNDAMENTAL PARTICLE INTERACTIONS II'**, *S. Ferrara and J. Ellis (eds)* *Plenum Press, New York and London (1983)*
vol.20.-Erice 1982-	**'GAUGE INTERACTIONS: Theory and experiment'**, *Plenum Press, New York and London (1983)*
vol.21.-Erice 1983-	**'HOW FAR ARE WE FROM THE GAUGE FORCES ?'**, *Plenum Press, New York and London (1984)*
vol.22.-Erice 1984-	**'QUARKS, LEPTONS, AND THEIR CONSTITUENTS'**, *Plenum Press, New York and London (1985)*
vol.22a.-Erice 1984-	**'FLAVOR MIXING IN WEAK INTERACTIONS'**, *Ling-Lie Chau (ed)* *Plenum Press, New York and London (1984)*
vol.23.-Erice 1985-	**'OLD AND NEW FORCES OF NATURE'**, *Plenum Press, New York and London (1986)*
vol.24.-Erice 1986-	**'THE SUPERWORLD I'**, *Plenum Press, New York and London (1987)*
vol.24a.-Erice 1986-	**'FUNDAMENTAL SYMMETRIES'**, *P. Bloch, P. Pavlopoulos and R. Klapisch (eds)* *Plenum Press, New York and London (1987)*
vol.25.-Erice 1987-	**'THE SUPERWORLD II'**, *Plenum Press, New York and London (1988)*
vol.25a.-Erice 1987-	**'SPECTROSCOPY OF LIGHT AND HEAVY QUARKS'**, *U. Gastaldi, R. Klapisch and F. Close (eds)* *Plenum Press, New York and London (1989)*
vol.26.-Erice 1988-	**'THE SUPERWORLD III'**, *Plenum Press, New York and London (1989)*
vol.26a.-Erice 1988-	**'HEAVY FLAVOURS AND HIGH-ENERGY COLLISIONS IN THE 1-100 TeV RANGE'**, *A. Ali and L. Cifarelli (eds)* *Plenum Press, New York and London (1989)*
vol.27.-Erice 1989-	**'THE CHALLENGING QUESTIONS'**, *Plenum Press, New York and London (1990)*
vol.27a.-Erice 1989-	**'HIGGS PARTICLE(S)'**, *A. Ali (ed)* *Plenum Press, New York and London (1990)*
vol.28.-Erice 1990-	**'PHYSICS UP TO 200 TeV'**, *Plenum Press, New York and London (1991)*

Enrico Fermi and Isidor I. Rabi, 1952, West Coast (from T.D. Lee Dedication Ceremony, Erice, 4 July 1996).

I am not an expert in Sicilian history, but out of this people I remember Archimedes, and Ettore Majorana, examples of men who made so much for the progress of physics. It is in this tradition that this whole institution is dedicated to the creation and propagation of high levels of knowledge.

Perhaps the fact that we talk one language in this School points towards a more hopeful future where it will perhaps be possible to recover the universalities which existed some centuries ago when latin was the universal language, when scholars travelled anywhere and wrote to one another in latin.

I would like now to make some reflection on the School as an outsider, but still a very keen outsider. This School is held in this very beautiful place which one would think is distracting. And it is distracting, but in a wonderful way, which enhances one's spiritual and intellectual sensibilities. The virtue of the School is in its utmost seriousness: students and professors come at 9 in the morning and work through very often until late at night, with serious discussions, in very serious exchanges of views and methods, and find criticism, which is also very important. Professor Zichichi runs in a very able way the period of questions, criticism and commentaries. At this School it is not just coming and listening to some lectures, but later on we shall have the opportunity to quiz the lecturer: and a good question is very highly regarded and rewarded. And this is much more than you can have in an ordinary classroom in the Universities or things of this sort. This is a real discussion, not always amongst equals, there are scientists of very distinguished achievements, but they provide a measure against which the younger people can prepare themselves and perhaps later achieve more than they otherwise would.

Isidor I. Rabi, from "New Phenomena in Subnuclear Physics", Plenum Press, New York and London, 1975, p. 8.

The author and Isidor I. Rabi, Geneva 1962.

vol.28a.-Erice 1990- **'MEDIUM-ENERGY ANTIPROTONS AND THE QUARK-GLUON STRUCTURE OF HADRONS',**
R. Landua, J.-M. Richard and R. Klapisch (eds)
Plenum Press, New York and London (1991)

vol.29.-Erice 1991- **'PHYSICS AT THE HIGHEST ENERGY AND LUMINOSITY: To Understand the Origin of Mass',**
Plenum Press, New York and London (1992)

vol.29a.-Erice 1991- **'QCD AT 200 TeV',**
L. Cifarelli and Y. Dokshitzer (eds)
Plenum Press, New York and London (1992)

vol.30.-Erice 1992- **'FROM SUPERSTRINGS TO THE REAL SUPERWORLD',**
World Scientific (1993)

vol.31.-Erice 1993- **'FROM SUPERSYMMETRY TO THE ORIGIN OF SPACE-TIME',**
World Scientific (1994)

vol.32.-Erice 1994- **'FROM SUPERSTRING TO PRESENT-DAY PHYSICS',**
World Scientific (1995)

vol.33.-Erice 1995- **'VACUUM AND VACUA: The Physics of Nothing',**
World Scientific (1996)

vol.34.-Erice 1996- **'EFFECTIVE THEORIES AND FUNDAMENTAL INTERACTIONS',**
World Scientific (1997)

vol.34a.-Erice 1996- **'UNIVERSALITY FEATURES IN MULTIHADRON PRODUCTION AND THE LEADING EFFECT'**
L. Cifarelli, A. Kaidalov and V.A. Khoze (eds)
World Scientific (1998)

vol.35.-Erice 1997- **'HIGHLIGHTS: 50 YEARS LATER'**
World Scientific (1998).

Paul Adrien Maurice Dirac at Erice (1982).

SUBJECT INDEX OF THE MAIN TOPICS
IN THE ERICE SUBNUCLEAR PHYSICS VOLUMES

ONE OF THE BEST LECTURES ON SUPERSYMMETRY, ERICE, 1981

INTRODUCTION TO SUPERSYMMETRY

Edward Witten[*]

Joseph Henry Laboratories
Princeton University
Princeton, New Jersey 08544

I. INTRODUCTION

Supersymmetry is a remarkable subject that has fascinated particle physicists since it was originally introduced.[1] Although supersymmetry is no longer a new idea, we still do not know in what form, if any, it plays a role in the proper description of nature.

The present status of supersymmetry might be compared very roughly to the status of non-Abelian gauge theories twenty years ago. It is a fascinating mathematical structure, and a reasonable extension of current ideas, but plagued with phenomenological difficulties.

In these lectures, I will present an introduction to supersymmetry, or at least to some aspects of this extensive subject. I also will describe some recent results. Supersymmetry has the reputation of a subject that is difficult to learn. I will try to at least partially dispel this unjustified impression.

In these lectures we will discuss only global supersymmetry, not supergravity.

[*] Supported in part by NSF Grant PHY80-19754.

305

The Unity of the Fundamental Interactions

PLENUM PRESS • NEW YORK AND LONDON

Eugene P. Wigner with Paul Adrien Maurice Dirac at Erice (1982).

FROM THE DIRAC MUSEUM

STATEMENTS BY
DISTINGUISHED ERICE LECTURERS

SELECTED BY NORMA SANCHEZ, DIRECTOR OF THE MUSEUM

THE BIRTH OF THE ERICE SCHOOL

Victor F. Weisskopf, the author and Sidney D. Drell in Erice, 26 May 1963.

Lord Patrick Maynard Stuart Blackett, Pyotr L. Kapitza, Paul Langevin, Lord Ernest Rutherford, Charles Thomson Rees Wilson outside Cavendish Laboratory (1929).

«We experimentalists are not like theorists: the originality of an idea is not for being printed in a paper, but for being shown in the implementation of an original experiment.»

Lord Patrick Maynard Stuart Blackett, 1962

Richard P. Feynman (1964).

«So in both theory and experiment, to improve the physics of the future, I would like to see a little more pride in the work.»

Richard P. Feynman, Erice 1964

Tsung Dao Lee at Erice, (1968).

M. Schwartz, Tsung Dao Lee, the author and Isidor Isaac Rabi at Erice (1968).

«What is now disproved was once thought self-evident.»

Tsung Dao Lee, Erice 1968

*Mrs Harriet Drell, Isidor Isaac Rabi and Mrs Maria Ludovica Zichichi
at the dinner to celebrate Rabi's 75th birthday, Erice (1972).*

«*Physics needs new ideas. But to have a new idea is a very difficult task: it does not mean to write a few lines in a paper. If you want to be the father of a new idea, you should fully devote your intellectual energy to understand all details and to work out the best way in order to put the new idea under experimental test.*

This can take years of work. You should not give up. If you believe that your new idea is a good one, you should work hard and never be afraid to reach the point where a new-comer can, with little effort, find the result you have been working, for so many years, to get.

The new-comer can never take away from you the privilege of having been the first to open a new field with your intelligence, imagination and hard work. Do not be afraid to encourage others to pursue your dream. If it becomes real, the community will never forget that you have been the first to open the field.»

Isidor Isaac Rabi, 1972

Sidney Coleman at Erice (1979).

«The Symmetries of the Vacuum are the Symmetries of Nature.»

Sidney Coleman, Erice 1973

FEYNMAN DIAGRAMS IN THE FEYNMAN LECTURE HALL
AT THE I.I. RABI INSTITUTE IN ERICE

Compton effect.
Elastic scattering
between an electron
and a photon. Two
basic processes contribute
to it.

The basic interaction
between electrons
consists in the exchange
of one (virtual)
photon.

Interaction of
an electron with
itself: an electron
emits a virtual
photon and absorbs
it again.

—— : electron
⌇⌇⌇ : virtual photon

Vacuum polarization effect.
A virtual pair electron-positron
can be formed and
annihilated again.

Each line in a Feynman diagram describes the path of a particle; when a particle breaks into two, its line divides as well. A mathematical expression is associated with each line and vertex in a Feynman diagram. The product of these expressions gives the amplitude of the probability that the depicted process occurs.

«There is a theory named after a fellow who has never written a paper on it. This reminds me of the show "How to succeed without even trying".»

Richard P. Feynman, Erice 1972

Laura Fermi at Erice (1975).

«Some physicists' wives believe that their husbands like physics better than they do their wives. And they may have a point. After work in the evening, when a wife is expecting a word of endearment, like "I couldn't live without you", the husband is most likely utterly silent, absorbed in scribbling numbers and symbols on the margins of the evening paper. When she would like to go to the movies, he has a date with an experiment that cannot wait. There are other complaints, some more justified than others. But all in all, life with a physicist is well worth living.»

Laura Fermi

from My Life as a Physicist's Wife, Erice, 16 July 1975, a Lecture in the series "The Glorious Days of Physics", at the International School of Subnuclear Physics.

Gabriele Veneziano at Erice (1998).

Gabriele   Veneziano at Erice (1998).

«*The Constants of the Vacuum are the Constants of Nature.*»

Gabriele Veneziano, Erice 1989

Gerardus 't Hooft at Erice (1998).

«Copying is easy, logical reasoning is difficult.»

Gerardus 't Hooft, Erice 1997

This statement of mine can easily be misinterpreted, so let me expand on it a little: Humanity has had thousands of years to learn how to apply past experiences in new situations. When some action, or method, has proved to be successful in the past, we apply a similar procedure to new situations. We can do this because we accumulated a miraculous ability at pattern recognition, which is what we call intelligence, and it is at the basis of the many successes of humanity as a biological species.

We successfully recognised numerous patterns in the physical world. But sometimes we encounter problems so novel that our past experiences appear to be of little help, and in fact may even mislead us. Nature often presents us with barriers that can only be overcome by meticulous and hard work. Looking back at what eventually gave us the required clues, I believe that it was "logical thinking". I do *not* mean "original thinking". Original thinking is actually easy. I (and other colleagues of mine) get letters every week from people with very original ideas, but, alas, their ideas are plainly wrong. It is correct logical thinking that is needed, and, here, I find that humanity has much less experience. In the past, copying — without true understanding — was a much faster route towards success than strict logic. You can learn how to use a bow and arrow without any understanding of mechanics. I do *not* want to say that copying is bad; we should use our past experiences wherever we can, but some of the barriers that we see ahead of us can only be taken by extremely rigorous logical reasoning, and, well, we are not good at that.

Gerardus 't Hooft

With his statement, 't Hooft worded his view on the progress made, and progress still to be made, in theoretical physics. In note [215], there is my testimony on the validity of the same point in experimental physics.

A prime example of 't Hooft's statement is the discovery of the Standard Model. It was only after this model was discovered that cross-sections, decay times and branching ratios could be calculated with extraordinary accuracy, and as experiments also reached a high degree of perfection, accurate comparison with experiment became possible. To achieve these unprecedented results, theoreticians had to overcome a notorious barrier in the 1960's: what are the rules for building detailed theories for ultrarelativistic quantum particles? Should we believe in quarks? Is there universality in the weak interactions? Why?

There were many false leads. Renormalization did not seem to be logically correct, but this impression was due to incomplete understanding. Now we have learned that a very careful treatment of the notions of unitarity and causality, using the mathematical formalisms of ghost particles in the virtual states, form the basis of a strictly logical approach that provides us with an extremely rigid framework. Not only the Standard Model resulted from that; the formalism led to the so-called BRST quantization procedure [167], on which all subsequent theoretical activities in quantum gravity, supergravity, superstring theories and their successors are based. Without knowing how to renormalize gauge theories, these subsequent developments could not have taken place. More details are in Note [216].

In conclusion, 't Hooft's work on renormalization of gauge theories, broken and unbroken ones, is a milestone on the road towards the Standard Model, as well as the theories beyond. More about the Standard Model in Note [217].